Biological Essentialism

Biological Essentialism

MICHAEL DEVITT

OXFORD
UNIVERSITY PRESS

OXFORD
UNIVERSITY PRESS

Great Clarendon Street, Oxford, OX2 6DP,
United Kingdom

Oxford University Press is a department of the University of Oxford.
It furthers the University's objective of excellence in research, scholarship,
and education by publishing worldwide. Oxford is a registered trade mark of
Oxford University Press in the UK and in certain other countries

First Edition published in 2023
Impression: 1

Published in the United States of America by Oxford University Press
198 Madison Avenue, New York, NY 10016, United States of America

British Library Cataloguing in Publication Data
Data available

Library of Congress Control Number: 2022947259

ISBN 978–0–19–884028–2

DOI: 10.1093/oso/9780198840282.001.0001

Printed and bound by
CPI Group (UK) Ltd, Croydon, CR0 4YY

For Pegg, a real beaut sheila

Contents

Preface

Although I have mostly worked in the philosophy of language, I have always been interested in the philosophy of biology. I would tend to go to talks on the subject when at philosophy conferences. Some of my best friends are philosophers of biology. But it was not until 2003 that I started working on the subject. I was prompted do so in writing a paper defending the linguistic thesis that the Kripkean notion of "rigidity" we need for kind terms is one of *rigid application* not one of *rigid designation* (2005).[1] The prompt came because this thesis, when applied to the likes of 'tiger', raised issues of biological essentialism. That led me to read a very instructive paper by Samir Okasha (2002) in which he set out the received views about essentialism in the philosophy of biology. These views struck me as quite wrong. For, they deny any intrinsic genetic component to the essence of a species or other biological taxon. And they implicitly deny that any member of a species is essentially a member.

So, without more ado, I wrote an eight-page piece I called, "Some Heretical Thoughts on Biological Essentialism". I sent this to every philosopher of biology I knew, and many I did not. This had two surprising consequences. First, the volume of response was astounding: initial responses together with follow-up discussions amounted to 100 pages. Second, given the consensus, I expected the experts to identify deep flaws in these "heretical thoughts". Yet this did not happen. I was corrected, informed, and guided on many matters, always in a wonderfully helpful way, and yet my basic argument for biological essentialism seemed to me to survive fairly intact. That argument was, in brief, that biological explanation demanded intrinsic essences. So, to the horror of some friends, I went seriously to work on these issues. This led to several publications, starting with "Resurrecting Biological Essentialism" (2008), and finally to this book.

While writing the book in 2020, some other related, and rather "hot", issues pressed in on me: issues of biological race "realism" and essentialism. I decided to include those issues in the book.

So, what are the issues that concern the book? Setting aside race for a moment, the issues are as follows:

1. What is it *to be* a member of a particular biological taxon? *In virtue of what* is an organism, say, a *Canis lupus*? What *makes* it one? These are various ways to ask

[1] I first presented this linguistic thesis in Devitt and Sterelny (1999: 85), largely stimulated by my anonymous reviewing of what was to become LaPorte (2000).

about the 'essence', 'nature', 'identity', even 'definition', of a particular taxon. They raise the issue of *taxon essentialism*.

2. What it is *to be* a particular individual organism? *In virtue of what* is an organism, say, the Queen? What *makes* it her? These are various ways to ask about the "essence", "nature", or "identity" of a particular individual. They raise the issue of *individual essentialism*.

3. If an individual organism belongs to a taxon does it do so essentially? This is the issue of *essential membership*. Clearly, if we had answers to both *taxon essentialism* and *individual essentialism* we would have an answer to *essential membership*: an organism *O* is essentially a member of a taxon *T* iff an organism having the essence of *O* entails its having the essence of *T*.

These essentialism issues have been much discussed by metaphysicians. Thus, on *taxon essentialism*, Saul Kripke (1980), Hilary Putnam (1975), and David Wiggins (1980) have proposed views that are similar to mine. My view is that the essence of a taxon, particularly a species, is (at least partly) an *intrinsic*, underlying, probably largely genetic property. This view accords with common sense and has been widely accepted in philosophy. These authors also embraced *essential membership*. And, talking about the Queen in particular, Kripke has urged a view on *individual essentialism*: her origin in certain gametes from certain parents is essential to her. This "origin essentialism" has stirred controversy among metaphysicians.

The methodology of the metaphysicians is to appeal to intuitions.

What have philosophers of biology had to say on these issues? The contrast with metaphysicians could hardly be more stark. First, philosophers of biology (and biologists) are dismissive of the popular Kripkean view on *taxon essentialism*. The idea that a species has an underlying intrinsic essence is thought to smack of "Aristotelian essentialism" and reflect a naive and uninformed view of biology that is incompatible with Darwinism. Clearly, if the essence of a species is not intrinsic it must be relational (assuming that it has an essence at all). The consensus is indeed that the essence is relational: for an organism to be a member of a certain species, it must have a certain *history*. Second, until recently, the issue of *essential membership* had been largely ignored in philosophy of biology. Insofar as it has been addressed it has been rejected. Third, the issue of *individual essentialism* has been totally ignored in philosophy of biology.

The methodology of philosophers of biology is to appeal to biological theory.

In "Resurrecting", I went along with the consensus in accepting, without argument, that there is an historical *component* to the essence of a taxon. However, I went sharply against the consensus, particularly over species, in arguing that there is *also* an underlying intrinsic component. So I sided with Kripke and the folk against the philosophers of biology. But I did so following the methodology of philosophers of biology: I appealed to biological explanations not intuitions. This book starts with a reprint of "Resurrecting" in Chapter 1.

Ernst Mayr made an important distinction between two problems about species: the "taxon" problem and the "category" problem. In virtue of what is an organism in the taxon *lion* and not *tiger*? That is an example of the taxon problem. In virtue of what is the taxon *lion* a species and not a subspecies or genus? That is an example of the category problem. This distinction is widely accepted but its significance is often overlooked in discussions of biological essentialism and racial realism. The distinction is crucial to my case for intrinsic essentialism in "Resurrecting", and to the discussions that follow in this book.

"Resurrecting" received detailed and interesting criticisms from several philosophers of biology: Robert Wilson et al. (2007), Matthew Barker (2010), Marc Ereshefsky (2010), Richard Richards (2010), Tim Lewens (2012), Sarah-Jane Leslie (2013), Matthew Slater (2013), and Marion Godman, Antonella Mallozzi, and David Papineau (Godman and Papineau 2020; Godman et al. 2020). Chapter 2 defends my intrinsic essentialism from these criticisms. In so doing I hope to strengthen the case for that essentialism.

The consensus view that the essence of a taxon is wholly relational raises two questions. (A) Why believe it? (B) What *precisely* is this wholly relational essence? The literature provides surprisingly little in the way of plausible answers, particularly to (B). Concerning (A), Chapter 3 presents an argument that there is at least an historical *component* to the essence. The chapter argues against such answers as I have been able to find to (B). It urges instead that the relevant history of a taxon is of organisms *of a certain intrinsic kind* evolving into organisms of *a certain other intrinsic kind*, until we reach the taxon in question. So, the historical component to the essence requires an intrinsic component. So, this view is another challenge to the consensus in the philosophy of biology.

So far, the concern has been all with taxa. In Chapter 4, the book turns to individuals. Whereas *essential membership* has been a topic of interest in metaphysics it has been largely ignored in philosophy of biology until quite recently, as LaPorte (1997) pointed out. He set about remedying this situation. Whereas, he charges, "essentialists have tended to be rather naïve on scientific matters", he aims to approach the issue "in the light of biological systematics" (p. 97). This approach leads him to reject *essential membership*. Some other philosophers of biology have since joined him in this; for example, Griffiths (1999), Okasha (2002), and Leslie (2013).

So, these philosophers of biology urge, from a biological basis, a view of what is *not* essential to an individual organism. But neither they nor, so far as I can discover, any other philosopher of biology or any biologist, seriously address the broader issue of *individual essentialism*, the issue of what *is* essential to the organism. It seems that this issue, much discussed by metaphysicians, has *entirely* escaped the attention of philosophers of biology. Chapter 4 argues that it deserves attention.

Chapter 4 presents an argument from the explanatory concerns of biology for the Kripkean view that an organism, like a taxon, has a partly intrinsic, partly historical, essence. This together with the book's view on *taxon essentialism* yield an argument for *essential membership*. The consensus is wrong again.

Essential membership has become topical because of a series of papers, mostly in *Biology and Philosophy*, beginning with the one by Alex Levine (2001). Levine rejects *essential membership* and so holds that that any organism is only contingently a member of its species. He finds this contingency in conflict with the common thesis in biology that any organism selected as the "type specimen" for a species is necessarily a member of that species. Levine expresses the conflict neatly: "*qua organism*, the type specimen belongs to its respective species contingently, while *qua type specimen*, it belongs necessarily" (p. 334). In embracing *essential membership* in Chapter 4, I reject Levine's *qua-organism* thesis. In Chapter 5, I argue against his *qua-type-specimen* thesis.

Finally, I turn to the lively field of the philosophy of race, a field that engages philosophers with backgrounds from biology to social theory. A major concern of the field is whether race is biologically "real", whether race "exists". A related concern is with what races are or, as I put it, with their essences or natures. In Chapter 6, I consider these issues from the perspective developed earlier in this book and in an article, "Natural Kinds and Biological Realisms" (2011c). I find the issue of "racial realism" unclear in its blurring of the earlier-emphasized crucial distinction between taxon and category issues; in this case, between the issue of alleged races and the issue of the category **Race**. Armed with this distinction, I argue that there are racial kinds, in some sense, that are indeed "in the realm of the biological". These kinds, like those thought to be part of the Linnaean hierarchy, have essences that are partly historical and partly underlying intrinsic properties. This racial realism does not, of course, endorse any theory of races, particularly not racist ones that have been used as instruments of discrimination and oppression.

My work on all these issues has led to several papers that have been delivered in talks and somctimes published. I have already mentioned "Resurrecting Biological Essentialism" (2008). It was delivered at many places around the world, starting with some universities in Australia in November 2005. Since this publication has been the subject of the criticisms discussed in Chapter 2, I thought it best to reprint it as Chapter 1, without any changes except a few additional footnotes. A preliminary version of the publication "Defending Intrinsic Biological Essentialism" (2021a) was delivered at a workshop in honor of Kim Sterelny's 60th birthday at Mystery Bay (NSW, Australia) in November 2010. The publication drew on a much longer version that was delivered at the University of Rijeka as part of a series of lectures in April 2017. Chapter 2 is a modified and further expanded version of that publication, including also a version of another publication, my response (2020: 441–9) to Marion Godman and David Papineau

(2020). "Historical Biological Essentialism" (2018a) was first delivered at the University of Sydney in April 2017. Chapter 3 is a modified version of the publication. "Individual Essentialism in Biology" (2018b), was first delivered at Macquarie University, Sydney, in November 2015. Chapter 4 is a modified and expanded version of the publication. A version of Chapter 5, "Type Specimens and Reference", was rejected by two journals. But this cloud had a silver lining: it gave me some helpful insight into likely objections, which I address. Chapter 6, "Racial Realism and Essentialism" was the basis for two lectures at the University of Rijeka in October 2021 and several later talks elsewhere. Working on it inspired a paper, "The Minimal Role of the Higher Categories in Biology" (2023).

I have received comments and advice from many over the years since I aired "Some Heretical Thoughts", including from those who commented on the papers that the book draws on. Here is my best, but probably inadequate, attempt to list those who have helped in one way or another: Matthew Barker, Alberto Cordero, Michael Dickson, John Dupré, Marc Ereshefsky, Peter Godfrey-Smith, Paul Griffiths, Adam Hochman, Tim Juvshik, Philip Kitcher, Joseph LaPorte, Michael Levin, Antonella Mallozzi, Raj Nanavati, Karen Neander, Samir Okasha, Makmiller Pedroso, Georges Rey, Stephen Schwartz, Stephen Stich, Iakovos Vasiliou, Joel Velasco, Denis Walsh, John Wilkins, Andrea Woody. Finally, thanks to the members of my classes on "Biological Essentialism" at the Graduate Center.

Michael Devitt

Hudson, NY
October, 2022

1
Resurrecting Biological Essentialism

"Essentialism about species is today a dead issue" (Sober 1980: 249)
"Folk essentialism is both false and fundamentally inconsistent with the Darwinian view of species" (Griffiths 2002: 72)

1.1 Introduction

The idea that biological natural kinds, particularly a species like dogs, have intrinsic underlying natures is intuitively appealing.[1] It has been shown to be widespread even among children (Keil 1989). It was endorsed by a great philosopher, Aristotle. Under the influence of the logical positivists, Popper (1950), Quine (1960), and others, it fell from philosophical favor in the twentieth century until revived by Saul Kripke (1980), Hilary Putnam (1975), and David Wiggins (1980). Many philosophers probably now take the view for granted. If so, they are right out of touch with biologists and, especially, philosophers of biology. For, the consensus among philosophers of biology, and a widespread view among biologists, is that this sort of "Aristotelian essentialism" is deeply wrong, reflecting "typological" thinking instead of the recommended "population" thinking (Sober 1980: 247–8). This essentialism is thought to arise from a naive and uninformed view of biology, indeed to be incompatible with Darwinism.[2] This view is nicely presented and argued for in a paper by Samir Okasha (2002). I shall take that as my main text. I shall defend intrinsic biological essentialism. I think that the children are right and the philosophers of biology, wrong.[3]

[1] First published in the *Philosophy of Science*, 75 (Devitt 2008). Reprinted in *Putting Metaphysics First: Essays on Metaphysics and Epistemology* (Devitt 2010) with some additional material in footnotes, identified by "[2009 addition]". Many of these additions remain in the present version. Some others have been added, identified as "[2022 addition]".

[2] Michael Ruse places Kripke, Putnam, and Wiggins "somewhere to the right of Aristotle" and talks of them showing "an almost proud ignorance of the organic world" (1987: 358n). John Dupré argues that the views of Putnam and Kripke are fatally divergent from "some actual biological facts and theories" (1981: 66). [2009 addition] The standard story is that biology was in the grip of classical essentialism until saved by Darwin. Polly Winsor (2006) argues persuasively that this story is a fabrication of Ernst Mayr's.

[3] This paper was prompted by writing another one defending the thesis that the notion of rigidity we need for kind terms is one of *rigid application* not one of *rigid designation* (Devitt 2005). The view that natural kind terms are rigid appliers has the metaphysical consequence that a member of a natural kind

Biological Essentialism. Michael Devitt, Oxford University Press. © Michael Devitt 2023.
DOI: 10.1093/oso/9780198840282.003.0001

I start by saying something about essentialism in general and about the essentialism I shall defend in particular.

A property *P* is *an essential property* of being an *F* iff anything is an *F* partly in virtue of having *P*. A property *P* is *the essence* of being an *F* iff anything is an *F* in virtue of having *P*. The essence of being *F* is the sum of its essential properties. Essences can be fully intrinsic; for example, the essence of being gold is having atomic number 79. Essences can be partly intrinsic and partly extrinsic and relational;[4] for example, the essence of being a pencil is partly being an instrument for writing, which an object has in virtue of its relation to human intentions, and partly having the sort of physical constitution that distinguishes it from a pen, which an object has intrinsically. Finally, essences can be fully relational and extrinsic; being Australian is probably an example because it seems that anything—Rupert Murdock, Phar Lap (a horse), the Sydney Opera House, a bottle of Penfolds' Grange, the expression "no worries mate", and so on—can have the property provided it stands in the right relation to Australia.[5]

The doctrine I want to defend, which I shall call "*Intrinsic Biological Essentialism*", abbreviated sometimes to "*Essentialism*", is that Linnaean taxa have essences that are partly intrinsic underlying properties. This calls for some clarification and comment.

(i) By "Linnaean taxa" I mean kinds that are thought to fall under the biological categories in the Linnaean hierarchy: kingdoms, phyla, classes, orders, families, genera, species, and even subspecies (varieties).[6] I do not mean kinds like those of predators or parasites. And I do not mean the categories themselves. *Essentialism* is a thesis about what it is for an organism to be, say, a dog not a cat, not about what it is for, say, dogs to be a species not a genus. (This distinction will loom large in sections 1.5 to 1.9.) The focus of my discussion will be on species but, I emphasize, *Essentialism* covers kinds that fall under all the categories.

is essentially a member. This sort of "individual essentialism" needs to be distinguished from the "kind essentialism" that is the concern of the present paper. [2022 addition] Individual essentialism is discussed in Chapter 4.

[4] Biological essentialism is usually taken to be concerned *only* with what is intrinsic (e.g., Mayr 1963: 16; Sober 1993: 146; Wilson 1999b: 188). This reflects the influence of Aristotle. I think it more helpful to define essentialism in a more general way so that issues come down to the *sort* of essence that a kind has.

[5] Locke called an underlying intrinsic essence that is causally responsible for the observable properties of its kind a "real essence". This is contrasted with a "nominal essence" which is picked out by reference-determining descriptions associated with a kind term. So, having atomic number 79 is the real essence of gold and the essence of being Australian, whatever it may be, is merely nominal. Kripke and Putnam showed that natural kind terms like 'gold' are not associated with reference-determining descriptions and so do not pick out nominal essences; they pick out real essences without describing them. This is not to say that a term *could not* pick out a nominal essence that is also real; indeed, 'having atomic number 79' is such a term (cf. Boyd 1999: 146).

[6] I say "thought to fall" because I sympathize with the doubts of some about this hierarchy; see Ereshefsky (1999; 2001); Mishler (1999). [2022 addition] There is a discussion of these doubts later (6.7).

(ii) I include the qualification "at least partly" because I shall not take issue with the consensus that a species is partly an historical entity.[7]

(iii) In sexual organisms the intrinsic underlying properties in question are to be found among the properties of zygotes; in asexual ones, among those of propagules and the like.[8] For most organisms the essential intrinsic properties are probably largely, although not entirely, genetic. Sometimes those properties may not be genetic at all but in "the architecture of chromosomes", "developmental programs" or whatever (Kitcher 1984: 123).[9] For convenience, I shall often write as if the essential intrinsic properties were simply genetic but I emphasize that my *Essentialism* is not committed to this.

(iv) *Intrinsic Biological Essentialism* would certainly be opposed by the consensus because of its commitment to *intrinsic* essences. But the consensus should not be opposed to biological essentialism in general because, as I am understanding essentialism, the consensus is that species have essences but these are *extrinsic or relational.* And Kim Sterelny and Paul Griffiths, in their excellent introduction to the philosophy of biology, *Sex and Death*, are explicitly not opposed to this sort of essentialism: "the essential properties that make a particular organism a platypus... are historical or relational" (1999: 186). Of course, the very term 'essentialism' has become so distasteful to biologists because of its association with Aristotelian metaphysics that a biologist would doubtless be reluctant to admit to any sort of essentialism. But the essentialism I have defined need not come with those Aristotelian trappings. Many philosophers would be similarly reluctant because the term 'essentialism' strikes them as quaintly old-fashioned, scholastic, even unscientific. But such reluctance would be a merely verbal matter. The issue of essentialism would remain even if the term 'essentialism' were dropped. It is the issue of in virtue of what an organism is a member of a certain Linnaean taxon; the issue of what makes an organism a member of that taxon; the issue of the very nature of the taxon. I stick with 'essentialism' because it is the

[7] However, I say that the essences are "at least, partly" intrinsic rather than simply "partly" because I do wonder whether *all* species are, or should be, partly historical. Citing the possibility of regularly produced hybrids like the lizard *Cnemidophorus tesselatus*, Philip Kitcher claims that "it is not necessary, and it may not even be true, that all species are historically connected" (1984: 117). [2022 addition] Historical essentialism is discussed in Chapter 3.

[8] What I would like is a term for asexual organisms that is like 'zygote' for sexual ones in referring to the beginning of an organism. John Wilkins informs me that there is no one term for this. Others he mentions include 'bud' and 'gemmule'. He has also drawn my attention to other uses of 'propagule'. Thus, consider the following definition:

> In animals, the minimum number of individuals of a species capable of colonizing a new area. This may be fertilized eggs, a mated female, a single male and a single female, or a whole group of organisms depending upon the biological and behavioral requirements of the species. In plants, a propagule is whatever structure functions to reproduce the species: a seed, spore, stem or root cutting, etc. http://www.radford.edu/~swoodwar%20/CLASSES/GEOG235/glossary.html

[9] Webster and Goodwin (1996) promote the idea of "morphogenetic fields".

term that philosophers of biology use for the doctrine that they want to reject and I want to promote. Those who are offended by the term should replace it with one of the other ways of characterizing the issue.

(v) There is some controversy over whether species are natural kinds or individuals. Michael Ghiselin (1974) and David Hull (1978) seem to see individualism as an antidote to essentialism.[10] But, as Okasha points out, "the issues about essentialism . . . do not depend on which view of the ontological status of species we favour" (2002: 193–4; see also Kitcher 2003: 137–40).[11] Thus, if a species is an individual rather than a kind, our essentialism issue for species becomes that of saying in virtue of what organisms are *parts of* a certain species, for example, the species *Canis familiaris*. And the consensus answer should be that it is entirely in virtue of the organisms' historical or relational properties, entirely "because they are part of the genealogical nexus" (Hull 1978: 309), whereas my *Essentialism*'s answer would be that it is partly in virtue of the organisms' intrinsic underlying properties.[12] Indeed the essentialism issue can be posed "nominalistically" in a way that is noncommittal on the ontological status of species: In virtue of what is an organism, say, a *Canis familiaris*? And that is how I *do* pose the issue in section 1.5. I mean to be neutral on the ontological issue but for convenience will mostly talk of species as if they were kinds.

(vi) *Essentialism* is primarily concerned with the natures of the *actual* groups identified by the folk and biologists for explanatory purposes. This "descriptive" issue needs to be distinguished from the "normative" issue of the natures of the groups that we *should* identify for explanatory purposes. Clearly, we might not be doing what we should be doing. However, I shall write as if we are. If we are not, then my *Essentialism* should be taken to cover the groups that we should be identifying for explanatory purposes as well.

In section 1.2, I give evidence that the consensus really is opposed to *Intrinsic Biological Essentialism*. In section 1.3, I argue for the doctrine: explanations in biology demand that there be essential intrinsic underlying properties. I turn then to objections. In section 1.4, I describe the standard relational views of species which, according to the consensus, make *Essentialism* untenable. In section 1.5, I emphasize a distinction which is crucial to showing that the consensus is wrong about this. The distinction is between two questions. (1) What is it to be a member of any group that happens to be a species? (2) What is it for a group to be a

[10] "*Individualism* about species is an idea with close links to antiessentialism, both conceptually and historically" (Griffiths 1999: 211).

[11] Richard Boyd goes so far as to say that the distinction between species being individuals or kinds "is almost just one of syntax" (1999: 164).

[12] Also, we should note, if *Canis familiaris* is an individual, we can ask about its *individual* essence just as we can about that of any individual (n. 3). And the consensus answer should be that its essence is its being constituted by organisms that share historical or relational properties whereas my *Essentialism*'s answer would be that those organisms must also share certain intrinsic underlying properties. [2022 addition] Individual essentialism is discussed in Chapter 4.

species? In sections 1.6 to 1.9, I argue that the relational views of species are, primarily at least, answers to question (2). *Essentialism*, in contrast, is an answer to (1). Indeed, these relational views can, mostly, be happily wedded to *Essentialism*. Even the influential phylogenetic-cladistic view can be wedded if it loses some implausible features.[13] How has the consensus got it so wrong? My tentative diagnosis is that the error has arisen from conflating questions (1) and (2), a conflation encouraged by some mistaken thoughts about conspecificity. Finally, in section 1.10, I accommodate some general features of Darwinianism, associated with variation and change, features that are thought to undermine *Essentialism*. The accommodation requires acceptance of some indeterminacy in what constitutes a biological kind. But we must all accept that, whatever our views of *Essentialism*.

If the arguments in sections 1.4 to 1.10 are right, the stated objections to an essentialist doctrine like *Intrinsic Biological Essentialism* fail. Perhaps there are some unstated objections that would succeed. And perhaps these could provide the basis for showing that the arguments in section 1.3 in favor of *Essentialism* are inadequate. Given the strength and longevity of the consensus in biology against intrinsic essentialism, it seems reasonable to predict this. Still, it remains to be seen whether it is so. At the very least I hope to show that the case for the consensus needs to be made a lot better than it has been.

1.2 Evidence of the Consensus

I have claimed that the consensus among philosophers of biology is that doctrines like *Intrinsic Biological Essentialism* are wrong. Among those philosophers, the claim hardly needs support because the consensus is so established. Still, among philosophers in general, the claim does need support because, influenced by Kripke and Putnam, many find the claim incredible and so think I must be struggling with a straw man. The epigraphs to this paper, drawn from the works of Elliott Sober and Paul Griffiths, two leading philosophers of biology, are some evidence that I am not. Here is some more.

The consensus starts by denying that members of a species share a distinctive set of genetic properties. Thus, according to Okasha,

> virtually all philosophers of biology agree that... it simply is not true that the groups of organisms that working biologists treat as con-specific share a set of common morphological, physiological or genetic traits which set them off from other species. (2002: 196)

[13] [2009 addition] Not so: the wedding is possible even with the implausible features; see nn. 43 and 44.

Clearly, if the members of a species do not share a distinctive set of genetic properties then those properties could not be essential properties of that species. Indeed, Okasha claims that "biologists and philosophers of biology typically regard essentialism about species as incompatible with modern Darwinian theory" (2002: 191). And John Dupré claims that "it is widely recognized that Darwin's theory of evolution rendered untenable the classical essentialist conception of species" (1999: 3). Alex Rosenberg says: "The proponents of contemporary species definitions are all agreed that species have no essence" (1985: 203). Mohan Matthen claims that "species . . . are associated with no nonrelational real essence" (1998: 115). Sober expresses this consensus as follows: "biologists do not think that species are defined in terms of phenotypic or genetic similarities"; tigers are "not *defined* by a set of traits" (1993: 148). Sterelny and Griffiths put the point bluntly: "no intrinsic genotypic or phenotypic property is essential to being a member of a species" (1999: 186). Ghiselin puts it even more bluntly: "That John Doe has a particular set of genes is about as relevant to his being a specimen of *Homo sapiens* L. as it is to his working for the manufacturers of Brand X" (1974: 283).

Finally, if the essence of a species is not in the least intrinsic then it must be entirely relational. I have already quoted Sterelny and Griffiths' claim to this effect about the platypus. And they think that nearly everyone agrees with them: there is "close to a consensus in thinking that species are identified by their histories" (1999: 8). Their view is endorsed by Okasha (2002: 202). Sober declares that tigers are "*historical entities*" (1993: 148). "Two organisms are conspecific in virtue of their historical connection to each other, not in virtue of their similarity" (1993: 150). Similarly, Marc Ereshefsky, speaking for "Darwin, the founders of the Modern Synthesis, and most cladists" (2001: 209). Finally, Hull claims: "If species are interpreted as historical entities, then particular organisms belong in a particular species because they are part of that genealogical nexus, not because they possess any essential traits. No species has an essence in this sense" (1978: 313). Ruth Millikan says much the same (2000: 19).

The consensus is broad but some are not part of it. Thus David B. Kitts and David J. Kitts (1979) urge an intrinsic essentialism like mine. According to Richard Boyd (1999) and Robert Wilson (1999b), species are "homeostatic cluster kinds" and I take it that they think that they have at least partly intrinsic essences.[14] And Philip Kitcher has this to say: "I want to remain agnostic on the issue of whether any species taxon has a nontrivial essence" (1984: 132, n. 16).

[14] Hilary Kornblith favors the view that species are homeostatic cluster kinds, notes that the members of the cluster need not be intrinsic, but does not take a stand on whether any of them are (1993: 111, n. 10). Griffiths writes approvingly of the homeostatic cluster view but argues that species have purely historical essences (1999: 217–22). [2022 addition] Wilson et al. include the property of *having a certain genotype* in the cluster (2007: 199). My Essentialism is clearly not a homeostatic cluster view.

I take the opposition to *Intrinsic Biological Essentialism* to be established. It is now time to argue for the doctrine.

1.3 An Argument for *Intrinsic Biological Essentialism*

I shall offer two reasons for believing *Essentialism*. The first is superficial but still, it seems to me, indicative of where the truth lies. Such essential properties seem to be part of what "genome projects" are discovering. The projects seem to be throwing light on the very nature of certain species. Thus the *New York Times* recently reported that researchers hope "to discover, from a three-way comparison of chimp, human and Neanderthal DNA, which genes have made humans human" (Wade 2006: A14).[15] Philosophers of biology disparage this common view (Sterelny and Griffiths 1999: 7; Okasha 2002: 197) but the view is certainly appealing.

The second reason is deep and shows why the view is appealing. We group organisms together under what seem, at least, to be the names of species or other taxa and make generalizations about the morphology, physiology, and behavior of the members of these groups: about what they look like, about what they eat, about where they live, about what they prey on and are prey to, about their signals, about their mating habits, and so on. These generalizations are the stuff of popular nature programs and are to be found throughout the writings of biologists and philosophers of biology. For example, we are told that ivy plants grow toward the sunlight (Sober 1993: 6); that polar bears have white fur; that Indian rhinoceri have one horn and African rhinoceri[16] have two (p. 21); that Hawaiian *Drosophila* "routinely form interspecific hybrids in the wild" (p. 156); that the Australasian bittern is superbly camouflaged (Sterelny and Griffiths 1999: 32); that "Major Mitchell cockatoos occasionally hybridize with galahs" (p. 189); that "Australian trees... are not just drought-proof; they are fireproof as well" (p. 203); that "magnetotactic bacteria... come equipped with little compasses called magnetosomes, which they use to navigate away from oxygen-rich surface water because oxygen is toxic to them" (p. 209).

Generalizations of this kind demand an explanation. *Why are they so?* Why, for example, is there this difference between the Indian and African rhinos? Such questions could, of course, be seeking an explanation of the evolutionary history that *led to* the generalization being true. Set that aside for a moment. The questions could also be seeking an explanation of *what makes* the generalization

[15] Consider also this recent news report in the *Scientific American* online: "'DNA barcodes are giving us a direct signal of where species boundaries lie,' says Paul Herbert, an evolutionary biologist at the University of Guelph in Ontario and a progenitor of the genetic bar code effort" (Biello 2007).

[16] [2022 addition] African rhinos are made up of two species, *Cereaotherium simum* ("White") and *Diceros bicornis* ("Black") and so are not strictly a Linnaean taxon.

true. Regardless of the history of its coming to be true, in virtue of what is it now true? What are the mechanisms? The truth of these generalizations cannot be brute facts about the world and so must be explained. Explanations will make some appeal to the environment[17] but they cannot appeal only to that. There has to be something about the very nature of the group—a group that appears to be a species or taxon of some other sort—that, given its environment, determines the truth of the generalization. That something is an intrinsic underlying, probably largely genetic, property that is part of the essence of the group. Indeed, what else could it be?[18] Some intrinsic underlying property of each Indian rhino causes it, in its environment, to grow just one horn. A different such property of each African rhino causes it, in its environment, to grow two horns. The intrinsic difference explains the physiological difference. If we put together each intrinsic underlying property that similarly explains a similar generalization about a species, then we have the intrinsic part of its essence.[19]

The generalizations we have been discussing reflect the fact that it is *informative* to know that an organism is a member of a certain species or other taxon: these classifications are "information stores" (Sterelny and Griffiths 1999: 195). But being a member of a certain taxon is more than informative, it is *explanatory*. Matthen points out that "many biologists seem committed to the idea that something is striped *because* it is a tiger" (1998: 115). And so they should be: the fact that an individual organism is a tiger, an Indian rhino, an ivy plant, or whatever, explains a whole lot about its morphology, physiology, and behavior. At first sight, the explanation of the animal's stripes may seem rather superficial, but it is not really. For, when biologists group organisms together under some name on the basis of observed similarities, they do so partly *on the assumption that those similarities are to be explained by some intrinsic underlying nature of the group*. It seems to me clear that this is their practice, whatever they say about essentialism.[20] So the apparently superficial explanation points to the deep fact that there is something intrinsic, probably unknown, partly in virtue of which the animal is a tiger and which causes it to be striped. That something is an essential intrinsic property. The sum of those properties, together perhaps with some historical ones, constitute the essence of a tiger. Sober rightly insists that the essence of a species

[17] The role of the environment is very obvious with plants. Thus the height of corn in a field depends on the temperature, the soil, and so on.

[18] The point is not, of course, that the explanation of any generalization, even any biological one, demands an intrinsic property, just that the explanation of a generalization of the kind illustrated demands one.

[19] So this intrinsic part is a *real* essence, in Lockean terms; see n. 5.

[20] The following comment of Sterelny is interesting in this respect: "Some, perhaps most, evolutionary biologists take speciation to occur only when there have been intrinsic changes". He finds this "puzzling for the view that species are historically defined entities is close to the consensus view in evolutionary biology". He is inclined to blame the influence of the folk who, as we noted (1.1), tend to be intrinsic essentialists (1999: 130). I think that the biologists and the folk are, deep down, tuned into the demands of explanation.

must explain why its members are the way they are. It must be "a causal mechanism that acts on each member of the species, making it the kind of thing that it is" (1980: 250).[21] That is exactly what this (partly) intrinsic essence is.

I distinguished two sorts of explanation that might be sought in asking why members of a species have a certain property. In so doing I am following in the footsteps of Ernst Mayr (1961). He regards an explanation of the mechanisms within members of a species that make a generalization true (regardless of the history) as concerned with "proximate" causation and part of "functional biology". In contrast an explanation of the evolutionary history that led to the mechanism being present in the members of a species is concerned with "ultimate" causation and is part of "evolutionary biology". The use of "ultimate" to describe the latter explanation seems like a gratuitous put-down of the former. Kitcher's even-handed description in response to Mayr's is *prima facie* more appropriate: "there are indeed two kinds of biological investigation that can be carried out relatively independently of one another, neither of which has priority over the other" (1984: 121). Adopting Kitcher's terminology, I shall call the former sort of explanation "structural" and the latter "historical". Structural explanations, as he says, seek to "explain the properties of organisms by means of underlying structures and mechanisms". He gives a nice example:

> A biologist may be concerned to understand how, in a particular group of bivalve mollusks, the hinge always comes to a particular form. The explanation that is sought will describe the developmental process of hinge formation, tracing the final morphology to a sequence of tissue or cellular interactions, perhaps even identifying the stages in ontogeny at which different genes are expressed. (p. 121)

He goes on to claim that "explanations of this type abound in biology: think of the mechanical accounts of normal (and abnormal) meiosis, of respiration and digestion, of details of physiological functioning in all kinds of plants and animals" (p. 121). And my point is that, at bottom, structural explanations will advert to essential intrinsic, probably largely genetic, properties.[22] It is because the bivalve mollusks have a certain intrinsic underlying nature that the hinge takes that form. That is the deep explanation.[23]

This discussion generates a number of questions. Here are two: (I) "Surely any of the generalizations we have been discussing could have exceptions: a small mutation may lead to an organism that seems to be a member of a species and

[21] [2022 addition] I later call this "the Sober demand" (2.5).

[22] Griffiths thinks not but his argument conflates structural explanations with historical explanations (1999: 210–11, 219–21).

[23] Hull unfavorably contrasts "classificationists" seeking "the unit of identification" with "phylogeneticists" seeking "the unit of evolution" (1965: 204). I think that the classificationists should be seen as seeking units of *structural explanation*, a very worthwhile pursuit.

yet lacks the property attributed to the species by a generalization. So the generalizations do not seem to be law-like. How does *Intrinsic Biological Essentialism* deal with that?"[24] (II) "It is of course the case that the truth of any such generalization must be explained by an intrinsic, probably largely genetic, property, but why does that property have to be an *essential* property of the kind in question?"[25] I shall consider these questions in section 1.10, along with others arising from Darwinian views of variation and change. I shall argue that *Essentialism* has an adequate answer to all these questions.

Sober claims that "evolutionary theory has removed *the need* for providing species with constituent definitions" and hence with intrinsic essences (1980: 255). I suspect that this sort of focus on evolution, hence on historical rather than structural explanations, has misled biologists and philosophers of biology about essentialism.[26] This having been said, I suspect that even historical explanations demand a partly intrinsic essence; that, for example, the explanation of how polar bears came to be white will ultimately depend on essential intrinsic properties of polar bears and of their grizzly ancestors. But I shall not attempt to argue this.

I have presented a positive argument for *Intrinsic Biological Essentialism*. We might sum it up: *structural explanations in biology demand that kinds have essential intrinsic properties*. That is my *first* main point in defense of *Essentialism*. In the rest of the paper I will develop the case for *Essentialism* in the course of responding to objections.

1.4 Relational Species Concepts

I start with what is alleged to be the central objection to *Intrinsic Biological Essentialism*: according to nearly all current "species concepts"—theories about the nature of species—species are *relational*. Okasha expresses this consensus as follows: "On all modern species concepts (except the phenetic), the property in virtue of which a particular organism belongs to one species rather than another is a relational rather than an intrinsic property of the organism" (2002: 201).

Despite the consensus that these species concepts make *Essentialism* untenable, the nature of biological species is, and always has been, an extremely controversial issue: "The species problem is one of the oldest controversies in natural history" (O'Hara 1993: 231); it is "one of the thorniest issues in theoretical biology" (Kitcher 2003: xii).[27] There are around two dozen species concepts and "at least

[24] "Philosophers of biology have often noted that there seem to be no laws which apply to all and only members of a species taxon (Hull 1978; Rosenberg 1985)" (Okasha 2002: 209).

[25] I am indebted to Peter Godfrey-Smith for raising this question.

[26] See, for example, Matthen (1998: 117–21), Griffiths (1999: 219–22), and Millikan (2000: 18–20).

[27] Although, interestingly enough, an issue that Darwin himself was skeptical about: he talks of "the vain search for the undiscovered and undiscoverable essence of the term species" (1859: 381).

seven well-accepted ones" (Ereshefsky 1998: 103). I shall follow Okasha in placing them in "four broad categories". In this section I shall briefly describe these concepts. In the next, I shall draw a distinction which is crucial to showing, in sections 1.6 to 1.9, that the consensus is wrong.

1. *Phenetic* concepts. On this sort of view, organisms are grouped into species on the basis of overall similarity of phenotypic traits. This is thought by its proponents to have the advantage of being fully "operational". Okasha says that phenetic concepts are "the least popular" (2002: 199) and this is hardly surprising because they arise from the "philosophical attitude... of empiricism" (Sokal and Crovello 1970: 29). "Phenetic taxonomists have often wanted to segregate taxonomy from theory" (Sterelny and Griffiths 1999: 196).[28] This category of species concept is the only one of the four that is *not* in the least historical and relational. I shall set it aside.

2. *Biological Species* concepts ("BSC"). The most famous example of BSC is due to Mayr. He defined species as "groups of interbreeding natural populations that are reproductively isolated from other such groups" (Mayr 1969: 26). Sterelny and Griffiths remark that "If the received view has a received species concept" it is BSC (1999: 188).[29]

3. *Ecological Niche* concepts ("ENC"). According to ENC, a species occupies a certain ecological niche. "A species is a lineage... which occupies an adaptive zone minimally different from that of any other lineage in its range and evolves separately from all lineages outside its range" (van Valen 1976: 70). Okasha puts the view succinctly: species "exploit the same set of environmental resources and habitats" (2002: 200).[30]

4. *Phylogenetic-Cladistic* concepts ("P-CC"). On this view we "identify species in terms of evolutionary history... [with] particular chunks of the genealogical nexus.... Species come into existence when an existing lineage splits into two... and go extinct when the lineage divides, or when all members of the species die" (Okasha 2002: 200). Sterelny and Griffiths claim that "something like a consensus has emerged in favor of a *cladistic* conception of systematics" (1999: 194). Nonetheless, it has some surprising features, as we shall see (1.9).

But perhaps the most important feature of the P-CC concept for the purposes of this paper is that it is, as everyone agrees, *incomplete*. It needs to be supplemented by a theory of speciation, a theory that explains when a lineage *has*

[28] Sterelny and Griffiths include under phenetic concepts those that define species in terms of *genetic* similarly (1999: 184). I clearly do not include these. I take the phenotype of an organism to be observable properties of it distinct from, but caused by, its genotype (along with the environment).

[29] Popular as it is, BSC has been the subject of extensive criticism; see, for example, Sokal and Crovello (1970); van Valen (1976); Cracraft (1983); Sober (1993: 155–6); Kitcher (1984: 118–20; 2003: 141–5); Mallet (1995); Dupré (1999); Sterelny and Griffiths (1999: 186–90). Mallet claims provocatively that the BSC concept "owes nothing either to genetics or to Darwinism" (p. 295).

[30] Ghiselin (1987: 374–8) has some severe criticisms of ENC.

split in two. For this, as Okasha says, P-CC "will have to rely on a concept of one of the other types" (2002: 201).

1.5 A Crucial Distinction

It is alleged that, according to each of these species concepts, except the phenetic which we are setting aside, species are relational. These concepts are thought, therefore, to show that the nature of a species could not be partly intrinsic and hence that doctrines like *Intrinsic Biological Essentialism* are false. In assessing this thought it is very important to distinguish the question that *Essentialism* is supposed to answer from another which it isn't.

Let *F*s be some group that has been classified for biological purposes under one of the taxa; for example, a group of poodles, dogs or *Canis*. The question that *Essentialism* answers has many forms, as already indicated (1.1, (iv)):

(1) In virtue of what is an organism an *F*?
What makes an organism an *F*?
What is the nature of being *F*?
What is the essence of being *F*?

This is a question about the *properties of organisms*. When it concerns *F*s that form a species, Mayr calls it the species "taxon" problem (1982: 253–4). So let us generalize this as "the taxon problem". It needs to be distinguished from the very different, "higher level", problem about the *properties of those properties*:

(2) In virtue of what are *F*s a subspecies, a species, a genus or etc.?
What makes a group of *F*s a subspecies, a species, a genus or etc.?
What is the nature of being a subspecies, a species, a genus or etc.?
What is the essence of being a subspecies, species, genus or etc.?

When it concerns species, Mayr calls it the species "category" problem (1982: 253–4). So let us generalize this as "the category problem".

"What is it to be a poodle not a bulldog?" is an instance of the taxon problem (1), "What is it for poodles to be a subspecies not a species?" is an instance of the category problem (2). The distinction between the two problems may seem obvious and yet it is easily conflated by certain forms of words. In particular, consider the question, "What is a species?" or "What is the nature/definition of a species?" These questions are ambiguous. They could be asking what sort of a nature any group has that happens to be a species, an instance of the taxon problem (1). But they are more likely to be asking what is it for any group to be a species, an instance of the category problem (2).

The distinction between the two problems is absolutely crucial to this paper. Which problem are the species concepts supposed to answer? According to Okasha, speaking for the consensus, they are at least supposed to answer problem (1) for species: they are supposed to show, as we noted, that "the property in virtue of which a particular organism belongs to one species rather than another is a relational rather than an intrinsic property of the organism" (2002: 201). Yet Ereshefksy, in his introduction to a collection that includes many classic papers on species concepts (1992a), says that "our concern is with a definition of the species category" rather than of the species taxon (1992b: xiv; see also Kitcher 1984: 120). And according to Sterelny and Griffiths (1999: 211) and Wilson (1999b: 191–2), the species concepts provide answers to *both* taxon and category problems. I shall argue that, on the contrary, the species concepts are primarily concerned with (2) and throw little light on (1).[31] Yet *Intrinsic Biological Essentialism* is concerned with (1). So, the central objection to *Essentialism*, based on the species concepts, fails. This is my *second* main point in defense of *Essentialism*.

In section 1.6 I shall show how much the species concepts bear on the category problem (2). In sections 1.7 and 1.8, I shall show how little they bear on the taxon problem (1).

1.6 Species Concepts and the Category Problem (2)

The species concepts straightforwardly answer problem (2) for species: they tell us about "the species category" (Sterelny and Griffiths 1999: 184).[32] Thus the popular biological species concept (BSC) tells us that a group is a species in virtue of being an interbreeding natural population that is reproductively isolated from other such groups. And the ecological niche concept (ENC) tells us that a group is a species in virtue of being a lineage which occupies an adaptive zone minimally different from that of any other lineage in its range and evolving separately from all lineages outside its range. And both these answers do indeed entail that being a species is *relational*: a group is a species in virtue of its breeding or niche relations to other groups.

The story for the influential phylogenetic-cladistic concept (P-CC) is a bit more complicated. As we have noted, the P-CC account of species adverts to the splitting of a lineage and so needs to be supplemented with a theory of splitting, a theory of speciation. Thus, according to P-CC, a group of organisms constitute

[31] Note that this is *not* the general claim that answers to (2) will throw little light on (1); indeed, see n. 33. It is a claim that *the relational species concepts* throw little light on (1).

[32] We are setting aside the phenetic concept but it is interesting to note that it is not obvious that this concept *does* answer (2). After all, subspecies like poodles and genera like *canis* could equally be identified by an overall similarity of phenotypic traits. What sort of similarity marks out species in particular?

one species at time t1 and their descendants constitute two daughter species at t2. But what makes it the case that the descendants are members of the daughter species rather than the original species? An account of speciation will tell us.

When faced with the need to supplement P-CC it is customary to wave a hand toward other species concepts to provide the needed theory of speciation. "The biological species concept, perhaps supplemented by the ecological species concept or by something else, reemerges as an account of speciation" (Sterelny and Griffiths 1999: 192). Thus, if P-CC is supplemented by BSC it will take a lineage to split when it yields two groups each of which is interbreeding but reproductively isolated from the other. And if it is supplemented by ENC, it will take a lineage to split when it yields two groups exploiting different sets of environmental resources and habitats. With some such supplement in mind, P-CC, just like the other species concepts, straightforwardly yields an answer to problem (2) for species. Thus, return to Okasha's statement of P-CC: "species come into existence when an existing lineage splits into two... and go extinct when the lineage divides, or when all members of the species die" (Okasha 2002: 200). Supplemented by a theory of speciation, this tells us what it is for a group of organisms to be a species rather than, say, a subspecies or a genus. And it tells us that this is a relational matter.

Not only do the species concepts straightforwardly yield answers to problem (2) for species, that seems to be what they are intended to do. Ever since Darwin, the species concepts have been tied closely to views of speciation and to distinguishing when two groups constitute subspecies of the one species and when they constitute two distinct species of a genus. They are concerned, for example, with whether the British red grouse (*Lagopus lagopus scoticus*) and its continental relative (*L. l. lagopus*) are separate species, and with whether the divergent forms of the bluegill sunfish (*Lepomis macrochirus*) constitute a single species. They are concerned with *what distinguishes species from other taxa*. And, we should note, they do not seem to be concerned with the taxon problem (1); with, for example, the nature of the British red grouse or the bluegill sunfish.[33]

Consider also the earlier-mentioned controversy over species (1.4). The arguments here are all over problem (2). Thus, alluding to the controversy, Ereshefsky says: "Biologists differ widely on how to define the species category... [on how] to provide the essential property of the species category – a property found in all and only species taxa" (1992b: xiv–xv). The controversy has led some (Mishler and Donoghue 1982; Kitcher 1984; Sterelny and Griffiths 1999: 194–201) to the view

[33] I take these examples from Mallet (1995). This paper starts with a nice discussion of the history of species concepts that makes their concern with (2) very apparent. Mallet himself urges a genotypic cluster definition of species as an answer to (2): "we see two species rather than one if there are two identifiable genotypic clusters. These clusters are recognized by a deficit of intermediates" (p. 296). This view puts him right outside what the philosophers of biology consider the consensus. Yet, he claims, "many, perhaps most, systematicists are currently using the genotypic (or morphological) cluster definition" (p. 298). Whether or not Mallet's answer to (2) is correct, it implies an answer to (1) that clearly is as congenial as could be to *Essentialism*.

that "there is no unique factor common to all species" (Ereshefsky 1992b: xv). This pluralism, as Robert Wilson notes, is "about the species category" (1999b: 192).[34] Biologists are concerned with whether groups that we have picked out for biological purposes should be counted as a subspecies, a species, a genus, or whatever. And with whether various considerations do, and should, play a role in settling such issues. In response to these issues, and inspired by cartographic generalization, Robert O'Hara urges that we take up the perspective of "systematic generalization" and then "we will be better able, not to solve the species problem, but rather to get over it" (1993: 232). It is quite clear that what he thinks we will be able to get over is fussing about when to judge a group to be a species. And, once again, the issues are not over problem (1), not over what is it for an organism to be a member of a group that we have picked out for biological purposes, irrespective of whether that group is a subspecies, species, genus, or whatever. So the issues are largely orthogonal to *Essentialism*. *Essentialism* is concerned with the nature of a group whatever the category it falls under.

I turn now to consider the bearing of species concepts on the taxon problem (1). In section 1.7, I will argue that BSC and ENC do not bear on this problem and so the consensus is wrong. In section 1.8, I will try to diagnose the error. In section 1.9, I will consider the bearing of P-CC.

1.7 BSC, ENC, and the Taxon Problem (1)

One reason for thinking that species concepts are not intended to answer the taxon problem (1) for species is that they quite obviously have nothing to say in answer to this problem for taxa other than species.[35] Yet if they were answering (1) for species we would expect them to be like *Essentialism* in saying *something*, at least, in answer to (1) for the other taxa, in saying *something* about what it is to be a member of a particular genus, for example.

Despite the consensus, neither BSC nor ENC provides an answer to (1). Let *F*s be a group of organisms that is a species according to BSC or ENC; for example, the group of tigers. What do BSC or ENC, as they stand, tell us about how to complete 'Some organism is an *F* in virtue of...'? Since *F*s are a species, BSC or ENC obviously tell us that whatever the completion it must specify *some property or other* of an organism that is at least *compatible* with the organism being a member of *some group or other* that, briefly, interbreeds or occupies a niche. But, beyond that, *they tell us nothing at all*! They don't tell us what property makes an

[34] [2022 addition] Pluralism has led to a "realism" issue about the **Species** category which I have discussed elsewhere (2009a; 2011c); see also later (4.8, 6.7.1).

[35] But the phenetic concept does answer the species taxon problem and could answer the taxon problem for other taxa; cf. n. 32.

organism a member of the group of *Fs in particular*. Indeed, they don't even tell us *what sort* of property that must be. As we have just noted, BSC and ENC tell us what it is for *Fs to be a species* rather than, say, a subspecies or genus, but they are silent on what it is for *an organism to be an F*, say, a tiger rather than a lion. Analogously, an account of what it is for a group of objects to be tools rather than, say, pets or toys would not tell us what it is for an object to be a hammer rather than, say, a saw. In brief, BSC and ENC, as they stand, say nothing about what *identifies* a particular species, hence nothing about what constitutes its essence. *The consensus view expressed by Okasha is simply false*: it is *not* the case that, according to BSC and ENC, "the property in virtue of which a particular organism belongs to one species rather than another is a relational rather than an intrinsic property of the organism" (2002: 201). These concepts, as they stand, say nothing about this matter and so need not be at odds with any doctrine that does. Indeed, they are not at odds with *Intrinsic Biological Essentialism*.

Essentialism's answer to taxon problem (1) is that the group *F* is identified (partly at least) by certain intrinsic underlying properties: it is in virtue of having such properties that an organism is a tiger rather than a lion. This is compatible with the BSC view that what makes *Fs* a species rather than a subspecies or a genus is a matter of interbreeding, and with the ENC view that it is a matter of occupying a niche. Indeed, it is better than compatible: *Essentialism partly explains* why the members of a species have the characteristics which, according to BSC or ENC, make them a species: it is partly *because* those members have certain essential intrinsic properties that, in the given environment, they interbreed and occupy a niche. Far from being undermined by these species concepts, *Essentialism* is complementary to them.

BSC and ENC do not *entail* a relational answer to (1) despite the consensus that they do. But perhaps we can see them as *associated with* such an answer. We can indeed find signs of two such answers in the literature. But neither could be a serious rival to *Essentialism*'s answer.

The first answer is the simple idea that what makes something an *F* is that its parents are *Fs*. This seems to be suggested by the following:

> the reference of an individual to a species is determined by its parentage, not by any morphological attribute. (G. C. D. Griffiths 1974: 102)
>
> If we suppose that humans first appeared about ½ million years ago, *Homo sapiens* is the name for the group that descended from the original organisms. (Ruse 1987: 344)

Hull quotes the Griffiths passage approvingly (1978: 305). He goes on to consider whether a human-like organism made by a scientist would *be* a human and this leads him to modify the simple idea: what makes something human is "being born of human beings *and/or mating with human beings*" (p. 306; emphasis added). A difficulty with the simple idea is that it rules out speciation: all organisms will be conspecific with their ancestors, however distant. Set that aside until later (1.10). The idea is open to an obvious objection: it is not really an answer to (1). It tells us that an organism is an *F* if its parents are *F*s. But what is it for *them* to be *F*s? The idea does not solve our problem, it simply moves it back a generation. And Hull's modification of the simple idea is open to a similar objection. It tells us that an organism is an *F* if it mates with *F*s. But what makes the organisms it mates with *F*s?

The second relational answer might be considered an elaboration of the first. It gets its inspiration from the typical naming practices of biologists: "Biologists coin new species terms by providing a *sample*, called a 'type specimen'" (LaPorte 2004: 5). Could we then identify a species by referring to its type specimen? So, what makes this interbreeding or niche-occupying group *F*s is that it contains a certain type specimen. This idea seems to be suggested by some other remarks of Hull:

> The taxonomist ... selects a specimen, any specimen, and names it ... A taxon has the name it has *in virtue of* the naming ceremony, not in virtue of any trait or traits it might have. (1978: 308)
>
> Any organism related to [the type specimen] in the appropriate ways belongs to its species, regardless of how aberrant the type specimen might turn out to be or how dissimilar other organisms may be. (pp. 311–12)

And consider this (entertaining) proposal for using BSC for "taxonomic definitions":

> Specify some individual, say Brigham Young, as your reference point, and then members of the same taxon are potential and actual interbreeders ...
>
> (Ruse 1987: 344)[36]

Combining this idea with BSC or ENC seems to suggest that what makes an organism a lion is that it is part of an interbreeding or niche-occupying group that contains a certain historically identified type specimen, say Leo; and what makes this other organism a tiger is that it is part of another interbreeding or niche-occupying group that contains a certain other historically identified type specimen, say Benji.

[36] Similarly, Matthen takes an organism to be a member of a particular species in virtue of belonging to the same extended reproductive community "as the originally ostended individual" (1998: 120).

This answer is transparently inadequate. Relating an organism to a type specimen may sometimes be a convenient way to *tell* what species the organism belongs to—for example, if the type specimen is held in some museum—but it should not be taken seriously as an account of what *constitutes* being a member of a species. The answer may be epistemically useful but it is *metaphysically hopeless*.[37] Why? Briefly, because being an *F*—for example, being a lion or being a tiger—is an *explanatory* property, as we noted in discussing the generalizations in section 1.3.

The hopeless answer itself immediately generates the demand for an explanation that it cannot possibly fulfill: *Why can lions interbreed with Leo and not Benji*? It is clearly no help to be told that that is what it is to be a lion. And this failure is just the tip of the iceberg. Consider the following questions (construed structurally not historically): Why do tigers have stripes? It is no help to be told that it is because they can interbreed with Benji. That does not tell us why any tiger, including Benji, has stripes. Why do polar bears have poor eyesight? Once again, their relation to some Ur-bear gives no explanation just as the relation of pieces of gold to the stuff in Fort Knox gives no explanation of why they are malleable. And so on through indefinitely many structural questions about the morphology, physiology, and behavior of species.

All these questions concern facts about species that could not be brute: the facts have to be explained. The suggested answers to (1) cannot provide adequate explanations. The moral of this discussion is that any adequate explanation cannot appeal only to relational properties of members of the species in question because those relations cannot bear the explanatory burden. An adequate explanation must appeal to intrinsic properties of the organisms. It is something about the intrinsic natures of lions, tigers, polar bears, and so on that provides the explanation (along with some environmental factors).

It is worth noting that the main point of the argument is not restricted to biology. Suppose we ask: Why do paperweights make good weapons? The answer is not that they are the same tool as a certain specimen paperweight kept in some museum. The answer is that the nature of paperweights makes them suitably heavy and easily grasped. And if they had a different nature, say that of buttons, then they would not make good weapons.[38]

So, contrary to consensus opinion, BSC and ENC do not give relational accounts of species identity. Indeed they do not give *any* account of species identity. Furthermore, they cannot be happily wedded to a relational account because such accounts are explanatorily hopeless. They can, however, be happily

[37] And for that reason it is charitable to construe any remark that seems to suggest this answer as making only an epistemic point.

[38] Despite this, I am not making any claim about explanation in general, just one about what is required for explanations of these phenomena in biology and similar ones elsewhere.

wedded to *Intrinsic Biological Essentialism* because intrinsic underlying properties can bear the explanatory burden.

We are left with a puzzle. BSC and ENC do not give a relational answer to the taxon problem (1) for species and yet the consensus is that they do. What has gone wrong? How can we diagnose the error?

1.8 The Conspecificity Route to Error about the Taxon Problem (1)

The obvious answer to the diagnostic question is that the error has come from somehow conflating the problems (1) and (2) that Mayr distinguished.[39] BSC and ENC do indeed offer relational accounts of what it is to be a species. But that is a very different matter from offering a relational account of what it is to be a member of a group that is a species.

Sterelny and Griffiths are interesting on this score. They nicely distinguish the taxon problem (1) from the category problem (2) in the "Further Reading" that concludes a chapter discussing the species concepts (1999: 211). Yet, surprisingly, the preceding discussion itself does *not* distinguish the problems (pp. 184–94). Indeed, they themselves draw attention to this conflation! Their defense is that "an answer to the taxon problem should solve the category problem, and vice versa" (p. 211).

This raises our puzzle in an acute form. *Why* would anyone think that an answer to the one problem would answer the other? The discussions in Sterelny and Griffiths, and in Okasha (2002), suggest that the route from a category answer to a taxon answer may be via an answer to the *conspecificity* problem, the problem of saying in virtue of what organisms are in the *same* species. This route is most explicit in Wilson's discussion (1999b). Wilson takes what are, in effect, the BSC and P-CC answers to the species category problem to imply answers to the taxon problem: they "imply that the properties determining species membership for a given organism are not intrinsic properties of the organism at all, but depend on the relations the organism bears to other organisms" (p. 192). How so? Because they imply that "conspecificity is not determined by shared intrinsic properties, but by organisms' standing in certain relations to one another" (p. 193). So the idea is that, (a), BSC and P-CC imply a relational nonintrinsic answer to the conspecificity problem and, (b), this implies a relational nonintrinsic answer to the taxon problem. And the problem with the idea is that, although (a) is clearly tempting, it is false. (b), however, is true. I shall start with it.

[39] Mayr's distinction is established but it is often overlooked: Dupré (1981); Stanford (1995) on which see Devitt (2009a); Griffiths (1999); Sterelny (1999).

The first thing to note in considering (b) is that an answer to the conspecificity problem does not alone provide an answer to the taxon problem. A conspecificity answer tells us what it is for two organisms to be members of the same species and hence what makes Leo not conspecific with Benji. A taxon answer tells us what it is to be a member of a particular species and hence what makes Leo a lion and Benji a tiger. These are two very different matters. An answer to the conspecificity problem does not answer the taxon problem because it does not determine *which* species conspecific organisms are members of; it does not determine species *identity*; it does not tell us that these conspecifics are lions, those, tigers. So, even if a species concept did answer the conspecificity problem, as (a) claims it does, more would still need to be done to answer the taxon problem. Still, the two problems are related in a way that sustains (b).

Suppose that the answer to the taxon problem is that an organism is a member of species *F* in virtue of being *Q*; and the answer to the conspecificity problem is that two organisms are conspecific in virtue of being *R*-related. Now, necessarily, if two organisms are both *Q* then they are both *F*s and hence conspecific. So the fact that they are both *Q* must *determine*, in a very strong way, that they are *R*-related and hence conspecific: *R* must *be* the relation of sharing a property of the *Q* sort, whatever that sort may be. Now suppose, as *Intrinsic Biological Essentialism* does, that *Q* is a partly intrinsic property of *F*s then *R* must be partly the relation of sharing that sort of intrinsic property. So if a species concept did imply a non-intrinsic answer to the conspecificity problem, as (a) claims it does, that would indeed count against *Essentialism*, as (b) claims.

Turn now to (a) and consider BSC. I have allowed that (a) is tempting. Yet BSC, as it stands, *says* nothing at all about conspecificity. So why is (a) tempting? Because, as Wilson notes, citing Mayr, BSC *implies* something about conspecificity: it implies that "a given individual organism is conspecific with organisms with which it can interbreed" (1999b: 192–3). And it is easy to think that this amounts to (a). But it doesn't, because BSC does *not* imply that organisms are conspecific *in virtue of* interbreeding. The crucial error is to suppose that it does imply this, to suppose, quoting Wilson again, that BSC implies that "conspecificity is... *determined* by... organisms' standing in certain relations to one another" (p. 193; emphasis added).[40] The point is a bit subtle and so I shall provide more details.

BSC, as it stands, straightforwardly provides an answer to the category problem. Now that answer implies that conspecific organisms are members of a group that is, as a matter of fact, an interbreeding (and reproductively isolated) group. For, to be conspecific is, by definition, to be members of a group that is a species and, according to BSC's category answer, what makes a group a species is being an

[40] Matthen provides another clear example of the error (1998: 117–21).

interbreeding group. The category answer tells us that conspecific organisms are members of an interbreeding group but it does not tell us *in virtue of what* they are members of that group. Indeed, BSC's category answer is compatible with *Intrinsic Biological Essentialism*'s answer to the conspecificity problem: it is compatible with the view that organisms are conspecific in virtue of sharing a certain intrinsic underlying property and, perhaps, a history. The compatibility is easy to see. *Essentialism* is motivated by the need to *explain* the observable properties of a group of organisms (1.3). These properties include, of course, the property of interbreeding. So, according to *Essentialism*, it is *because* the members of a species share the intrinsic underlying properties necessary to make them conspecific that, in the given environment, those members interbreed and hence have the property than makes them a species according to BSC.

Now we could, of course, *supplement* BSC as it stands with a relational answer to the conspecificity problem: organisms are conspecific *in virtue of* being able to interbreed. But, first, this supplement is not entailed by the BSC answer to the category problem and gets no support from the considerations that motivate that answer. Without further argument, the supplement is gratuitous. And, second, the supplement is a very bad answer to the conspecificity problem. It is a very bad answer *because*, as (b) shows, it is at odds with *Intrinsic Biological Essentialism*. (b) shows that a relational conspecificity answer entails a relational taxon answer and our earlier argument shows that relational taxon answers are explanatorily hopeless. The supplement is not something that BSC should be saddled with.

Although (a) is certainly tempting it remains puzzling that people would give in to the temptation: it remains puzzling that they would not have doubts about this conspecificity route to anti-*Essentialism*. For, although the answer to the conspecificity problem that BSC is wrongly thought to give does entail that the answer to the taxon problem must be relational and nonintrinsic it does not *give* such an answer. And as soon as one tries to give one, it should become apparent how explanatorily inadequate a relational answer must be. At least, this should be apparent if one keeps in mind the needs of structural explanations. So, it remains puzzling that a person who starts with the idea that BSC implies a relational conspecificity answer would not be led to contemplate one of two responses: abandoning BSC *simply because* of that implication; or, abandoning the idea that BSC has the implication.[41]

Our discussion of the puzzling conflation of the category problem with the taxon problem has focused on BSC. Yet the discussion applies just as much to ENC.

[41] Could the lack of doubt arise from conflating the false view that organisms are conspecific in virtue of being able to interbreed with the true view (assuming BSC) that they are conspecific in virtue of that in virtue of which they are able to interbreed? (Thanks to Michael Dickson.)

This concludes our discussion of how little bearing BSC and ENC have on the taxon problem (1). We turn now to consider the bearing of P-CC on that problem.

1.9 P-CC and the Taxon Problem (1)

At first sight P-CC, unlike BSC and ENC, may seem to provide an answer to the taxon problem (1) for species. Let *F*s be a group of organisms that is a species according to P-CC. Whereas BSC and ENC told us little about how to complete 'Something is an *F* in virtue of . . . ,' P-CC may seem to tell us a lot. The idea would be that since *F*s are a species, an organism is an *F* in virtue of being a member of a group with a particular "evolutionary history"; the species is "this particular chunk of the genealogical nexus". What makes this organism a lion not a tiger is that it is a member of a species having this particular place in "the branching tree-of-life" (Okasha 2002: 200). But this idea amounts to identifying a species simply by its relation to other species, to the species that preceded and succeeded it in its lineage: "You and I are members of *Homo Sapiens*, therefore, because we both belong to the segment of the genealogical nexus which originated in Africa some 300,000 [years] ago (on current estimates), and which has not budded off any daughter species since that point" (pp. 200–1). But this answer to (1) is inadequate in just the same sort of way as our earlier second answer in discussing BSC and ENC: its relational identification of a species is explanatorily hopeless. To see this, it helps if we drop the actual names of species (like "*Homo Sapiens*"), which might provide some illicit information, and replace them with schematic names using which we can capture the relational information that is all we are entitled to on this P-CC view. Thus suppose that species *A* splits into species *B* and *C*, then *B* splits into *D* and *E* and *C* splits into *F* and *G*. This is represented in the "tree-of-life" as depicted in Figure 1:

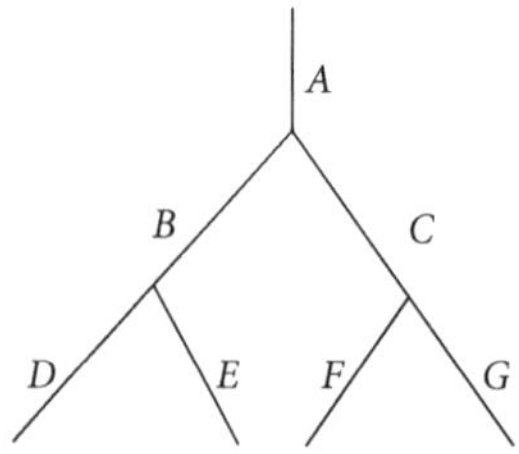

Figure 1

What does the P-CC view tell us about the nature of *B*? That *B* is descended from *A* and that *B* is distinct from *C*. And that's what P-CC tells us that about the nature of *C* too. So that clearly does not distinguish *B* from *C*. No more does it distinguish *D* and *E* to know that each is descended from *B*, and *F* and *G* to know that each is descended from *C*. Furthermore, since P-CC does not distinguish *B* from *C* it does nothing to distinguish *D* and *E*, descended from *B*, from *F* and *G*, descended from *C*. Suppose *D* were lions and *G*, tigers. Relational facts of the sort captured in this representation, which are the only facts that P-CC allows to constitute the natures of species, would do nothing to distinguish lions from tigers, hence nothing to explain the morphological, physiological, and behavioral differences between them. Suppose that we want to explain why *C* has poor eyesight, all we could appeal to on this view is its relation to *A*, *F*, and *G*. This is no more helpful in explaining the poor eyesight of *C* than was the relation of polar bears to some Ur-bear in explaining theirs. The structural explanations we need must appeal to the intrinsic properties of a species.

In sum, if P-CC is taken to give an answer to the taxon problem (1) for species, its answer is a very poor one. But it would be more charitable to suppose that it does not really intend to give an answer. Indeed, why would anyone think that it does? We have already noted a possible explanation in section 1.8: the route from a category answer to a taxon answer may be via an answer to the *conspecificity* problem, the problem of saying in virtue of what organisms are conspecific.

We have agreed with Wilson that a relational answer to the conspecificity problem implies a relational answer to the taxon problem; that is (b) in section 1.8. Wilson also thinks that P-CC, like BSC, implies such an answer to the conspecificity problem: P-CC implies that "conspecificity is determined by a shared pattern of ancestry and descent" or by something similar (1999b: 193); that is (a) in section 1.8. Now P-CC does indeed seem to give such a relational answer to the conspecificity problem, just as it seems to give a relational answer to the taxon problem. But that conspecificity answer would be bad because that taxon answer would be explanatorily hopeless, as we have seen. So, if P-CC really did involve such an answer it should be abandoned. But it is more charitable to suppose that, despite appearances, P-CC does not really propose an answer to the conspecificity problem.

In section 1.8 I argued that the account given by *Intrinsic Biological Essentialism* of what it is for an organism to be a member of a particular species—an answer to (1)—can be wedded happily to the accounts that BSC and ENC give of what it is for a group to be a species—answers to (2). Can it also be wedded happily to the influential P-CC? No. Adjustments would have to be made.

There is no problem wedding *Essentialism* to P-CC's view that species are historical entities because *Essentialism* is not committed to a fully intrinsic essence. The wedding would yield the view that a species is constituted partly by intrinsic, probably genetic, properties and partly by a particular history: an organism is a member of a certain species *F* in virtue of having certain intrinsic properties and being part of a particular chunk of the genealogical nexus. Those intrinsic properties are the ones that, together with environmental properties, explain the morphological, physiological, and behavioral properties of members of *F*.

The chunk of the genealogical nexus in question should be the one that plays a role in explaining the evolution of the species.

But this wedding of *Intrinsic Biological Essentialism* to P-CC could not include two rather surprising features of P-CC. (i) P-CC does not allow anagenesis, the forming of a new species without any split in the old. No matter how dramatically a lineage changes it will not form a new species unless it splits (Hennig 1966; cf. Simpson 1945). So if *Homo sapiens* had evolved from protists without any splits, all the organisms in this lineage would be in the same species. Kitcher aptly notes that "this strikes many people as counterintuitive (even insane)" (2003: 151). *Essentialism* could not go along with this rejection of anagenesis (assuming, as we obviously do, that some groups covered by *Essentialism are* species).[42] For, as Sterelny and Griffiths point out, the rejection of anagenesis has the consequence that "there is no upper limit to the amount of evolutionary change that can take place within one species" (1999: 7). So there would be no limit to genetic variation in a species and hence its essence could not consist of genetic properties.[43] But the rejection of anagenesis is deeply implausible. Why? Because a doctrine that, in some possible world, places *Homo sapiens* and protists in the same species seems to have lost track of the explanatory significance of an organism being a *Homo sapiens* or a protist.

(ii) Suppose that a species *A* splits off a side branch that forms a daughter species *B* but the population otherwise remains unchanged; that is to say, apart from the members of *B*, the descendants of the members of *A* do not differ phenotypically or genetically from their ancestors. So, had there not been the split forming *B*, those descendants would have all been members of *A*, on anyone's view of species. Yet P-CC has the consequence that because *B* did come into existence those descendants form a new species *C*. *Essentialism* cannot accept this. According to *Essentialism*, if a population remains unchanged then its members must be conspecifics.[44] But the P-CC view is another implausible one, and for a

[42] It is an interesting empirical question whether there are many plausible actual cases of anagenesis.

[43] [2009 addition] The claim that the rejection of anagenesis is incompatible with *Essentialism* is mistaken. Call the group of organisms constituting the lineage from protists to the final descendants of *Homo sapiens*, "*PH*". According to P-CC, *PH* would be a species if there had been no splitting in the lineage. However implausible this is it is no problem for *Essentialism* because *Essentialism* has nothing to say about when a group is a species. *Essentialism* is simply committed to the view that if *PH* were a species (or any other Linnaean category) then it would have a partly intrinsic essence. And *PH would* have such an essence: there would be something intrinsically in common to all those organisms, although probably not very much. The truth underlying the mistaken claim, and the problem for P-CC, is that *PH* is not a group worth identifying for the purposes of structural explanations, hence not a group worth identifying as a species in any circumstances. Its essence is not explanatorily interesting enough.

[44] [2009 addition] This claim is also mistaken. Given the story, it follows from *Essentialism* that *A* and *C* share the intrinsic part of their essences. Of course, the two groups differ in their histories: *A* is the ancestor of *B* but *C* is not. According to P-CC this makes *A* and *C* different species. But this is quite compatible with *Essentialism* because, once again, *Essentialism* has nothing to say about when a group is a species. The truth underlying the mistaken claim, and the problem for P-CC, is that *A* and *C* are not groups worth distinguishing for the purposes of structural explanations.

closely related reason. If a population remains unchanged then its members should all be grouped together in explaining their common morphological, physiological, and behavioral properties. That's what the explanatory role of species seems to require.[45]

In sum, *Essentialism* could be wedded to P-CC at the cost of dropping P-CC's two surprising, and implausible, features.

This concludes the discussion of the bearing of species concepts on the taxon problem (1) for species. Despite the consensus, BSC and ENC do not give a relational answer to that problem and if P-CC is taken to do so, its answer is a very poor one. My tentative diagnosis of the error in the consensus is that it arises from a conflation of the taxon problem with the category problem, a conflation encouraged by some mistaken thoughts about conspecificity. Perhaps also a focus on evolution has taken attention away from the needs of structural rather than historical explanations.

1.10 Variation and Change

The consensus among philosophers of biology is that doctrines like *Intrinsic Biological Essentialism* are at odds with Darwinian evolutionary theory (Sober 1980; Griffiths 2002). We have been discussing what is thought to be the most important problem for such doctrines: contemporary historical views of species. We must now consider a number of other alleged problems centering on variation and change.

Variation

(A) Sober claims that "no genotypic characteristic can be postulated as a species essence; the genetic variability found in sexual populations is prodigious" (1980: 272). Others write in the same vein. Thus Wilson rejects genetic essentialism because:

> The inherent biological variability or *heterogeneity* of species with respect to both morphology and genetic composition is, after all, a cornerstone of the idea of evolution by natural selection. (1999b: 190)

[45] Sterelny and Griffiths claim that, according to P-CC, the levels of the traditional Linnaean hierarchy above the base level of species "make little sense" (1999: 201). If this were taken as a view of the *taxa* then *Essentialism* should not go along with it. But P-CC does not support such a view of the taxa. The claim should be taken rather as a view of the *categories*, in which case it is quite compatible with *Essentialism* (Devitt 2009a; 2011c).

And Okasha claims:

> Intra-specific genetic variation is extremely wide – meiosis, genetic recombination and random mutation together ensure an almost unlimited variety in the range of possible genotypes that the members of a sexually reproducing species can exemplify. It simply is not true that there is some common genetic property which all members of a given species share and which all members of other species lack. (2002: 196)

Surely, one thinks, this must be an exaggeration. Surely there are genetic properties that humans share and that say chimpanzees, let alone worms, lack. Indeed, even Mayr, no friend of essentialism, talks of "the historically evolved genetic program shared by all members of the species" (1963: 17).[46] And Okasha goes on: "This is not to deny, of course, that there are important genetic similarities between members of a single species... species taxa are distinguished by clusters of covarying [chromosomal and genetic] traits, not by shared essences" (2002: 197). Great! So the clusters *are* the essences!! On the strength of these remarks, it seems as if the consensus should be simply that the crude idea that there is, say, "a tiger gene" is wrong. But to reject that crudity is not to reject the idea that a certain cluster or pattern of underlying, largely genetic, properties is common and peculiar to tigers.[47] So my *third* main point in defense of *Intrinsic Biological Essentialism* is: *an intrinsic essence does not have to be "neat and tidy"*. And, because the intrinsic essence is identified by its causal work, we need not be concerned that the identification will be ad hoc: the essence of the Indian rhino is the underlying property that does, as a matter of fact, explain its single horn and other phenotypical features.

(B) Okasha emphasizes the importance of variation to natural selection: "Darwinianism leads us to expect variation with respect to organismic traits, morphological, physiological, behavioural and genetic. For genetically-based phenotypic variation is essential to the operation of natural selection" (2002: 197). Sober thinks that this variation clashes with essentialism's commitment to the Aristotelian "Natural State Model": essentialism takes the variation to be the result of "interfering forces" taking an organism away from its "natural state" (1980: 257–9); to be "the result of imperfect manifestations of the idea implicit in each species" (Mayr 1963: 16); to be "deviation" from an "ideal" (Griffiths 2002: 78–9). This contrasts with the Darwinian view: "Individual differences are not *the effects* of interfering forces confounding the expression of a prototype; rather they

[46] Kitcher (1984: 132–3, n. 27) refers to other similar suggestions in the literature. And note this claim by Hebert, as reported in the earlier-cited item from the *Scientific American* online (n. 15): "We have very low levels of variation within a species and this deep divergence between species" (Biello 2007).

[47] The evidence seems to point to genes that switch other genes on and off—for example, Hox genes—being particularly important to the nature of a biological kind (Carroll 2005).

are *the causes* of events that are absolutely central to the history of evolution" (Sober 1980: 264). Furthermore, "the Natural State Model presupposes that there is some phenotype which is the natural one *which is independent of a choice of environment*" (Sober 1980: 268).

Essentialism need not go along with the teleological thinking of the Aristotelian Model and *Intrinsic Biological Essentialism* does not.[48] That doctrine can and should accept the Darwinian view of variation: variation within a species is indeed to be expected; species are indeed, as Griffiths says, "pools of variation" (2002: 78). *Essentialism* is committed simply to the view that in the pool of variation among the members of a species there are shared intrinsic, probably largely genetic, properties. And *Essentialism* rejects the idea that it is "not natural" for a corn plant of a particular genotype to wither and die, owing to the absence of trace elements in the soil (cf. Sober 1980: 268).

Gradual Change

Hull puts the problem for essentialism thus: "according to evolutionary theory, species develop gradually, changing one into another. If species evolved so gradually, they cannot be delimited by means of a single property or set of properties" (1965: 203; see also Ruse 1987: 347). According to Griffiths essentialism "is precisely the 'typological' perspective on species that Darwin had to displace in order to establish the gradual transformation of one species into another" (2002: 77; see also Caplan 1980: 73).[49] But Darwin didn't have to. Suppose that *S1* and *S2* are distinct species, on everyone's view of species, and that *S2* evolved from *S1* by natural selection. *Essentialism* requires that there be an intrinsic essence *G1* for *S1* and *G2* for *S2*. *G1* and *G2* will be different but will have a lot in common. This picture is quite compatible with the Darwinian view that the evolution of *S2* is a gradual process of natural selection operating on genetic variation among the members of *S1*. Indeed, gradual change is obviously compatible with having essential intrinsic properties: rivers, mountains, continents, planets, and so on, are all the result of gradual change and yet all have partly intrinsic natures.

Still, there may seem to be a worry, nicely expressed by Sober:

> evolution is a gradual process. If species *A* gradually evolves into species *B*, where in this lineage should one draw the line that marks where *A* ends and *B* begins? Any line will be arbitrary. Essentialism, it is alleged, requires precise and non-arbitrary boundaries between natural kinds... (1993: 147)

[48] Nor need it go along with Hull's "three essentialist tenets of typology" (1965: 201).

[49] Similarly Ereshefsky, writing about the essentialism of Lyell and Lamarck, claims that "their conception of species as evolving entities conflicts with this essentialist requirement" (1992b: xv).

This raises three issues: indeterminacy (or vagueness), arbitrariness, and "worldmaking".

Indeterminacy

Ereshefsky, paraphrasing Hull (1965), starkly puts the problem that indeterminacy is alleged to pose: "The boundaries of species are vague... there is no genetic or phenotypic trait that marks the boundary from one species to the next. Therefore no trait is essential for membership within a species" (1992c: 188–9). But this is a mistake: *Essentialism* does *not* require sharp boundaries between species. On the *Essentialist* picture, the evolution of *S2* from *S1* will involve a gradual process of moving from organisms that determinately have *G1* to organisms that determinately have *G2* via a whole lot of organisms that do not determinately have either. There is no fact of the matter about where precisely the line should be drawn between what constitutes *G1* and what constitutes *G2*, hence no fact of the matter about where precisely to draw the line between being a member of *S1* and being a member of *S2*. Essences are a bit indeterminate.

There are two reasons not to be worried by this. First, indeterminacy is everywhere. It is indeterminate whether a certain *x* is a mountain, or a certain *y*, a planet,[50] but this does not show that that there is no essence to being a mountain or a planet. Mount Everest has the somewhat indeterminate essence of being a mountain and yet is determinately a mountain; Mars has the somewhat indeterminate essence of being a planet and yet is determinately a planet. Second, there is just the same level of indeterminacy about species *whatever one's (Darwinian) view of them and of essentialism*, as indeed the passage from Sober indicates. For, *everyone agrees* that there comes a point where two organisms that have some common ancestor are nonetheless of different species. Yet there is no determinate matter of fact about precisely where that point is. And it is very easy to spot the root of the problem. We are tempted to say that an offspring and its parent are conspecific whatever the mutation (as Okasha indicates; 2002: 197). But if we do say this, it is obvious that *all* organisms will (probably) come out conspecific. Biology faces a classic *sorites* problem. *The indeterminacy that biology must learn to live with is no special problem for Intrinsic Biological Essentialism.*[51] That is my *fourth* main point in the defense of the doctrine.

We have been talking about the issue of indeterminacy in the world. This is likely to raise a worry about arbitrariness. There may indeed be some arbitrariness

[50] The recent debate by the International Astronomical Union shows that Pluto is a good example of this indeterminacy.

[51] "Essentialism is in principle consistent with *vague essences*" (Sober 1980: 253). Sober also draws attention to the fact that Aristotle was aware of "line-drawing problems" (pp. 252–3).

in the way we talk about the world. But that is not to say that there is arbitrariness in the world.

Arbitrariness

Biologists choose, for various explanatory purposes, to introduce names for certain groups of organisms thought to be species. There could be some arbitrariness about what groups to choose. Let us start with the worst case. Suppose that we allow for anagenesis, as my *Essentialism* must (1.9):[52] a new species can be formed without any split in a lineage. Suppose next that evolution were not only gradual but also *steady*: the morphological, physiological, behavioral, and genetic properties of organisms in the lineage change at a steady rate. *Essentialism* alone does not rule this out. How then would we choose where to draw our (indeterminate) lines in naming the species of this lineage? Clearly, there would be a deal of arbitrariness about this choice.[53] But we should not exaggerate how much. Our explanatory purposes in introducing a name for a species demand that we draw the lines around a group that is small enough to share a whole lot of important properties and large enough to yield broad generalizations. That is what is required for structural explanations. And, as G. G. Simpson points out, "such arbitrary subdivision does not necessarily produce taxa that are either 'unreal' or 'unnatural'" (1961: 60–1). Furthermore, *wherever we draw the lines* in naming a group "*F*", it is still the case that the intrinsic essence of being *F*, together with the environment, explains the morphological, physiological, and behavioral properties typical of *F*s.

Turn next to the best case. This is the situation if the hypothesis of *punctuated equilibrium* (Eldredge and Gould 1972) is right. On this hypothesis, evolution is far from steady. Species do not change much over most of their existence and then, in a relatively short period of time, either go extinct or evolve into daughter species. So, on my *Essentialist* picture, the need to explain the morphological, physiological, and behavioral properties of organisms would dictate that organisms in the period of stasis formed a species that should be named. We would draw our (indeterminate) lines in the period of rapid change. Our choice would hardly be arbitrary at all.

The important point for our purposes is that, wherever the truth lies between the worst and the best cases, arbitrariness poses no threat to *Essentialism*. The groups we name will still have partly intrinsic essences. Indeed, arbitrariness is really a problem for the species *category* rather than the taxa.

[52] [2009 addition] Not so: see n. 43.

[53] "The idea then is that if phenotypic change does not proceed by large jumps (*saltations*), then species are not objectively identifiable over time" (Sterelny and Griffiths 1999: 180. They do not endorse this idea.).

"Worldmaking"

We *do* have a choice about what groups of organisms to *name* "*F*". It is sadly common to confuse this with a choice we certainly *do not* have: the choice about which things *are F*. This is, in effect, the distinction between *making theories* and *making worlds*, a distinction the importance of which can hardly be exaggerated.[54] We name a group of organisms "*F*" for explanatory purposes and hence, even at worst, the choice of what group to name is mostly not arbitrary. But, *however arbitrary it is*, indeed even if it was *totally* arbitrary, we would not thereby make those organisms *F*. When biologists chose to apply the name '*Drosophila melanogaster*' to a vast number of insects, they did not thereby *make* those insects *Drosophila melanogaster*. They always were *Drosophila melanogaster* and would have been even if there had been no biologists around to call them anything.[55] It is common to talk as if, in doing science, we impose our concepts to "divide up reality". But this is not literally so: we choose our concepts in an attempt to discover the causally significant features of a nature that is already "divided up".[56]

Monsters

This discussion provides the wherewithal to deal with "monsters", offspring that differ greatly from their parents. Monsters are thought to refute *Essentialism* because they lack what might plausibly be proposed as the intrinsic essences of their parents' species. For example, Okasha claims that "if a member of the species produced an offspring which lacked one of the [essential] characteristics, say because of a mutation, it would be very likely to be classed as con-specific with its parents" (2002: 197). Sterelny and Griffiths put the point more firmly: "No intrinsic genotypic or phenotypic property is essential to being a member of a species.... People born with the wrong number of chromosomes, eyes, or arms are still human beings. So the essential properties that make a particular organism a platypus, for example, are historical or relational" (1999: 186). Now Okasha is surely right that we would very likely classify any offspring as conspecific with its

[54] I have argued this at length elsewhere (1997, particularly ch. 13; 2001). Overlooking the distinction seems to rest on something like a use/mention confusion.

[55] Kyle Stanford (1995) has a different view; see Ereshefsky (1998) and Devitt (2009a) for criticisms.

[56] It is easy even for staunch realists to slip into loose ways of talking that suggest worldmaking. Thus Kornblith says that when we "group objects together under a single heading on the basis of a number of easily observable characteristics... we thereby create a nominal kind" (1993: 41). But we don't! We create a *concept* that picks out a kind that may or may not be "real" in Locke's terms (n. 5 above) but which has its members independently of our creation. And Boyd, talking of kinds with nominal essences, says that their "boundaries" are "purely matters of convention" (1999: 142). But they aren't! Our naming a kind picked out by a certain set of descriptions is conventional but the boundary of the kind thus picked out is not.

parents. But the *sorites* problem shows that we cannot always be *right* to do so, whatever we think of *Essentialism*: as Hull says, "Obviously... there must have been instances in which non-horses (or borderline horses) gave rise to horses" (1978: 306). So what should the *Essentialist* say about monsters? One of two things. (a) If the mutations are gross enough, we should indeed say that the offspring is not of the same species as its parents. And that surely is what we would say, as monster movies sometimes illustrate. I doubt that we would even hesitate to say it of embryos that are so monstrous that they would not grow into viable organisms and are spontaneously aborted. (b) In other circumstances we should say that the status of the offspring is indeterminate. Return to our schematic example of the evolution of *S2* from *S1*. At the beginning of that process, there were organisms that determinately had *G1* and so were determinately members of *S1*, and at the end, there were organisms that determinately had *G2* and so were determinately members of *S2*. But in between there were organisms that did not determinately have *G1* or determinately have *G2* and so were not determinately members of *S1* or determinately members of *S2*. All we can say is that the further an organism gets from determinately having *G1* to determinately having *G2*, the further it gets from being determinately a member of *S1*. This is vague of course, but that's the way a lot of the world is, not just living things. Monsters are no special problem for *Essentialism*.[57]

Laws: We are now in the position to respond to the questions raised in section 1.3 about my treatment of biological generalizations. Question (I) was: "Surely any such generalization could have exceptions: a small mutation may lead to an organism that seems to be a member of a species and yet lacks the property attributed to the species by a generalization. So the generalizations do not seem to be law-like. How does *Intrinsic Biological Essentialism* deal with that?" *Essentialism* surely does demand that these generalizations be law-like rather than accidental. In a group of animals, it does not just *happen* to be the case that the members of a certain subgroup have one horn and the members of another, two. It is *because* the first subgroup are Indian rhinos and the latter, African rhinos; it is part of their very natures to have (in their environments) one

[57] Philip Kitcher, in commenting on an early version of this paper, claimed that "knockout" mutants produced by modifying "normal" zygotes, show that my treatment of monsters is too quick. It seems to me that my discussion accommodates these knockouts well enough: a minor mutant of a *Drosophila melanogaster* may still count as a *Drosophila melanogaster* because it has the essential intrinsic property that explains the characteristics it shares with "normal" *Drosophila melanogaster*; a gross mutant would not count as a *Drosophila melanogaster* because it does not have that property; the status of other mutants, doubtless most of the mutants, is simply indeterminate. We can learn about *Drosophila melanogaster* from these mutants, as we did, even if they themselves are not determinately *Drosophila melanogaster*. There is plenty of room for subtlety here. And if I am right in my arguments, something along the lines of my proposal has no viable alternative.

horn and two horns, respectively. But how can the generalizations be law-like if there could be exceptions?[58]

There are several things we might say in answer. First it is common, perhaps even the rule, for laws in the special sciences to have exceptions: they hold only *ceteris paribus*. So why should this be a problem for biology in particular? Indeed, if Nancy Cartright (1983) is right the situation is not much different in physics. Second, *statistical* generalizations can be law-like. Thus the claim that, say, 90% of *F*s are *P* can be law-like: it can sustain the subjunctive conditional that if something were an *F* it would very likely be *P*.[59] Finally, we can say that universal biological generalizations are indeed law-like but that there is some indeterminacy about precisely which organisms they would cover. 'All *F*s are *P*' may be a law in that anything that would be determinately *F* would be *P* but there might be some organisms that would not be determinately *F* or determinately not *F* and so there would be no determinate matter of fact about whether the law covered them. Note that this is not primarily an epistemological problem of *telling* what organisms the generalizations cover: it is primarily a metaphysical problem. Of course, even where there is a determinate matter of fact that generalizations cover certain organisms there can still be a problem discovering this; thus, many black birds in Australia were determinately swans at a time when biologists believed that all swans were white.[60]

Essentialism?

Question (II) was: "It is of course the case that the truth of any such generalization must be explained by an intrinsic, probably largely genetic, property, but why does that property have to be an *essential* property of the kind in question?" Suppose that the generalization is 'All *F*s are *P*' and that the explanatory intrinsic property is *G*. So it is agreed that, ultimately, it is because *F*s have *G* that they have *P*. The question asks why we must take *G* to be an essential property of *F*s. My answer

[58] Note that exceptions that arise from varying the environment are not a problem. Indeed, typical generalizations about an organism are implicitly restricted to its "normal" environment.

[59] Griffiths points out that "the generalizations of the special sciences often fail to live up to the ideal of a universal exceptionless law of nature.... Nevertheless...they have '*counterfactual force*'" (Griffiths 1999: 216). Referring to history, social sciences, geology and meteorology, Boyd notes that "causally sustained regularities...need not be eternal, exceptionless, or spatiotemporally universal" (1999: 152). [2009 addition] Marc Lange (1995) has argued persuasively that statements of the form 'The *S* is *T*' about a species *S* can express natural laws even though there could be *S*s that are not *T*. [2022 addition] Michael Thompson (2008) has a similar conclusion.

[60] [2009 addition]. We need to distinguish two sorts of potential "exceptions" to a law-like 'All *F*s are *P*': those arising from a mutation and those not. The objection seems to concern the former sort. My third, and final, response addresses this sort, denying that there can be such exceptions. Organisms that appear to be exceptions are not determinately *F*. But there can be exceptions that do not arise from mutations. My first and second responses address this sort.

rests on the just-argued claim that the generalization is law-like. So, anything that *would be* determinately an *F* would be *P* (in the appropriate environment). But now, in virtue of what is that the case? The answer is that anything that would be an *F* would have *G*. Indeed what other answer could we seriously entertain given that having *G* explains why all actual *F*s are *P*? We have now answered question (II). For, if anything that would be an *F* would have *G* then having *G* is essential to being an *F*: that is what it is to be an essential property.[61]

1.11 Conclusion

I have proposed the doctrine, *Intrinsic Biological Essentialism*: Linnaean taxa have essences that are partly underlying, probably largely genetic, intrinsic properties. The consensus in biology and philosophy of biology is that any such essentialism is deeply mistaken. In section 1.2, I set out evidence that this is indeed the consensus.

In section 1.3, I presented my central argument for *Essentialism*: the ubiquitous generalizations of biology need structural explanations that rest on essential intrinsic underlying properties of kinds. That was my *first* main point in defense of *Essentialism*.

In section 1.4, I described current species concepts. The consensus view is that these make doctrines like *Essentialism* untenable because, according to these concepts, species are *relational*. In section 1.5, I emphasized a distinction that is crucial to my defense of *Essentialism* from this consensus view. It is the distinction between two problems, a taxon problem (1) and a category problem (2): (1) What is the essence of being *F* (where *F*s are a group under one of the biological taxa)? (2) What is the essence of being a subspecies, species, genus, or etc.? This distinction yields two ways to understand the question "What is a species?" The question could be asking about the nature of any group that happens to be a species or it could be asking about what it is to be a species. My *second* main point in defense of *Essentialism*, argued for in the next two sections, was that the (partly) relational species concepts are primarily concerned with (2) whereas *Essentialism* is concerned with (1).

In section 1.6, I argued that, not only do the species concepts straightforwardly yield answers to question (2) for species, that seems to be what they are intended to do. In section 1.7, I argued that, contrary to the consensus, the biological species concept and the ecological niche concept do not answer (1) nor are they even associated with relational answers that are close to being explanatorily adequate. Indeed they can both be happily wedded to *Essentialism*'s (partly) nonrelational answer. Where has the consensus gone wrong? My tentative diagnosis in section 1.8

[61] [2022 addition] I have abandoned this answer to question (II) and now urge another; see section 2.5.

was that the error arises from a conflation of the taxon problem with the category problem, a conflation encouraged by some mistaken thoughts about conspecificity.

In section 1.9, I considered the influential phylogenetic-cladistic concept (P-CC). P-CC might be taken to give a relational answer to (1) as well as (2) but, if it were, its answer would be explanatorily inadequate. Can *Essentialism*'s answer to (1) be wedded to P-CC's answer to (2)? *Essentialism* can easily accommodate P-CC's view that species are historical entities. However, it cannot accommodate two of P-CC's features: its rejection of anagenesis and its view that a species must go extinct when it has a daughter.[62] But those features seem unwelcome anyway.

Finally, in section 1.10, I argued that some general features of Darwinism do not undermine *Essentialism*. Variation within a species can be seen to be compatible with *Essentialism* once one realizes that an intrinsic essence does not have to be "neat and tidy"—my *third* main point in defense of *Essentialism*—and that *Essentialism* is not wedded to the Aristotelian "Natural State Model". *Essentialism* can accept the gradual change of one species into another. Still, there are some concerns raised by the lack of sharp boundaries between species. First, *Essentialism* must accept a certain indeterminacy about species. But this is no worry because this indeterminacy has to be accepted whatever one's (Darwinian) view of species and of essentialism; biology faces a *sorites* problem. That was my *fourth* main point in defense of *Essentialism*. Next, *Essentialism* is compatible with there being a certain amount of arbitrariness in choosing which groups of organisms to name as species. But this choice, however arbitrary, must not be confused with a choice we do not have: the choice to make things members of a group we have named. The fact of indeterminacy enables *Essentialism* to deal with the problem of monsters—organisms lacking what might plausibly be proposed as the intrinsic essences of their parent's species—and to maintain the law-like status of biological generalizations despite apparent exceptions.

I have dealt with the objections to *Intrinsic Biological Essentialism* that I have found in the literature. Perhaps there are other objections that would be more effective. Perhaps it can be shown that my argument in favor of *Essentialism*—the argument from explanation—is inadequate. Given the strength and longevity of the consensus against such an essentialist doctrine, it seems reasonable to predict this. Still, it remains to be seen whether it is so. At the very least I hope to have shown that the case for the consensus needs to be made a great deal better than it has been.

If the arguments of this paper are good, the consensus relational view about particular species is quite mistaken. And many claims that biologists make day in and day out about the living world require species to have natures that they do not have according to this consensus.

[62] [2009 addition] Not so: *Essentialism* can accommodate these implausible features of P-CC; see nn. 43 and 44.

2
Defending *Partly Intrinsic Taxon Essentialism*

2.1 Introduction

What is the "essence", "nature", "identity", or "definition" of a biological taxon?[1] In "Resurrecting Biological Essentialism" (2008), Chapter 1 above, I answered that the essence consists partly in intrinsic underlying, and probably largely genetic, properties. I called this doctrine "*Intrinsic Biological Essentialism*" but, given the need to distinguish this doctrine clearly from other doctrines to come in later chapters, I shall rename it:

> *Partly Intrinsic Taxon Essentialism*: Biological taxa have essences that are partly intrinsic underlying, probably largely genetic, properties.[2]

Although something like *Partly Intrinsic* (for short) has been urged by some influential philosophers, Saul Kripke (1980), Hilary Putnam (1975), and David Wiggins (1980), and appeals to commonsense,[3] the consensus in biology, especially in the philosophy of biology, is that the answer is quite wrong, smacking of "Aristotelian essentialism", and reflecting a naive and uninformed view of biology

[1] This chapter modifies and greatly expands responses to my critics in Devitt (2021a) and Devitt (2020: 441–9). In both those works *Partly Intrinsic Taxon Essentialism* was called "Intrinsic Biological Essentialism", abbreviated "IBE".

[2] Here are three other recent defenses of something like *Partly Intrinsic*. (1) Denis Walsh (2006) gives a very interesting and subtle argument that intrinsic natures are needed to secure "the stability and mutability of individual organisms required for adaptive evolution" (p. 436). And he concludes, provocatively, that these natures are *Aristotelian* essences (p. 444)! (2) Travis Dumsday (2012) speaks approvingly of the arguments given by Walsh and me for intrinsic essentialism but goes on to give some arguments of his own. He starts with a wonderfully simple and persuasive argument for a *minimal* doctrine: in brief, *being an organism* is an intrinsic property of an organism and that property is essential to the organism. He thinks that this doctrine "can plausibly be thought to be presupposed by biologists" (p. 492). Perhaps so but, to my knowledge, it has never been acknowledged by members of the consensus in the philosophy of biology. I am less persuaded by Dumsday's subsequent argument for a stronger doctrine. (3) The biologist, Stuart Newman (2020), argues for intrinsic essentialism from "a 'physico-genetic' perspective that emphasizes the material properties of cell assemblages and their inherent properties" (p. 129). He sees his essentialism as differing from mine in rejecting "an exclusively gene-centric notion of intrinsic identity of organismal categories, focusing rather on the material properties of animal tissues (which have, of course, a genetic dimension)" (p. 143). For some other rejections of the consensus, see Oderberg (2007), Boulter (2012), and Austin (2017; 2019).

[3] It seems to appeal also to Robert Plomin, a professor of behavioral genetics at King's College London, as reported in the *Wall Street Journal*:

Biological Essentialism. Michael Devitt, Oxford University Press. © Michael Devitt 2023.
DOI: 10.1093/oso/9780198840282.003.0002

that is incompatible with Darwinism. I quoted Samir Okasha, who captures the consensus well:

> virtually all philosophers of biology agree that . . . it simply is not true that the groups of organisms that working biologists treat as con-specific share a set of common morphological, physiological or genetic traits which set them off from other species. (2002: 196)

On this matter, according to Sarah-Jane Leslie, "there is a degree of consensus among philosophers of biology (and indeed biologists) that is almost unprecedented in philosophy at large" (2013: 132).

Clearly, if a species has an essence at all and that essence is not intrinsic, then it must be relational. The consensus is indeed that the essence is relational: for an organism to be a member of a certain species, it must have a certain *history*. As Kim Sterelny and Paul Griffiths put it, there is "close to a consensus in thinking that species are identified by their histories" (1999: 8).[4] I went along with the consensus in allowing that there may be an historical *component* to the essence of a taxon (2008: 346; 1.1 above). I have since argued that there is (2018a, and Chapter 3 below). So my difference with the consensus, particularly over species, lies in my arguing that there is *also* an intrinsic underlying component: I argued for *Partly Intrinsic*.

This argument has received detailed and interesting criticisms from a number of philosophers: Robert Wilson, Matthew Barker, and Ingo Brigandt (2007), Matthew Barker (2010), Marc Ereshefsky (2010), Richard Richards (2010), Tim Lewens (2012), Sarah-Jane Leslie (2013), Matthew Slater (2013), and Marion Godman, Antonella Mallozzi, and David Papineau (Godman and Papineau 2020; Godman et al. 2020) This chapter is a response. My aim is to advance the positive case for *Partly Intrinsic* as well as defend the case already made in Chapter 1.

2.2 Clarifications

What are the taxa in question? They are those that are thought to fall under the biological categories (or ranks) in the Linnaean hierarchy of kingdoms, phyla, classes, orders, families, genera, species, and even subspecies. I say "thought to fall" because I sympathize with the doubts of some about this hierarchy (see, for

> About 99% of the 6 billion steps in the spiral staircase of DNA's double helix are the same for all of us. *This is what makes us human.* Behavioral geneticists are interested in the 1% of DNA *that makes us individuals.* (November 17, 2018: C1; emphasis added)

[4] Okasha endorses this view of the consensus (2002: 202). For some others, see Hull ([1978] 1992); Sober (1993); Matthen (1998); Millikan (2000); Ereshefsky (2001); LaPorte (2004).

example, Ereshefsky 1999, 2001; Mishler 1999). These doubts are often expressed as being about whether a certain Linnaean category—for example, **Genus**—"exists" or is "real".[5] This strikes me as a most infelicitous way of expressing the doubts. I argue that what is really being doubted is whether the category is *explanatory* (2011c). But whether or not the categories (or ranks) are explanatory, and hence have a proper place in biological theories, it is quite clear which taxa are thought to fall under them. Those are the groups that concern our essentialism issue. So the concern is *not* with some other biological groups; for example, *predators*, *parasites*, *females* (2008: 346; 1.1 above).[6]

There is an important further point: whatever the case with categories, our working assumption should be that these taxa, thought to fall under them, *are* very largely explanatory. Biologists, like anybody else, can identify and name any group they choose. But, when doing biology, they are obviously striving for an explanatory classification. And we should surely assume that they have mostly achieved this. We know from the history of taxonomy that mistakes have been made and they are surely still being made. Nonetheless, we should assume that our taxonomy is very largely good and explanatory. Thus, we should assume that *Canis* and *Canis familiaris* are explanatory notions whether or not their being, respectively, about a **Genus** and a **Species** are explanatory. We are concerned with the essentialism of these presumed explanatory taxa. We need take no position on whether they are properly placed in the Linnaean hierarchy of categories.[7]

The essences that we are concerned with are *property* or *kind* essences. A property *P* is *the essence* of being an *F* iff anything is an *F* in virtue of having *P*. In "Resurrecting", I said that the essence of being *F* is "the sum of its essential properties" (2008: 345; 1.1 above), but it might have been better to say that to have the essence of being *F* is to have any essential property of being *F*. Essences can be fully intrinsic (being gold); partly intrinsic and partly extrinsic and relational (being a pencil); or, fully relational and extrinsic (being Australian).

Some may resist any talk of "essence", thinking that the term smacks of Aristotelian metaphysics and has a scholastic air. But the term does not matter. What I am picking out with that term—also 'nature' and 'identity'—is the property *in virtue of which* an object is member of a certain kind; the property that *constitutes* its being a member or *makes* it a member. Those who find my terms for this property distasteful should choose another (2008: 347–8; 1.1 above). Some may think that there *is* no such property. I think that they are very wrong, for a reason I shall give in section 2.5.

[5] For example, Ereshefsky (1998: 113); Eldredge and Cracraft (1980: 327); Sterelny and Griffiths (1999: 197). Doubts about **Race** are expressed similarly; see sections 6.2 and 6.3 below.

[6] And the discussion may not apply to microbial kinds; see Ereshefsky and Reydon (2015) and Godman and Papineau (2020).

[7] But I do take a stand elsewhere (2023; and in section 6.7 below, discussing racial realism).

To say that kinds have essences (natures) is not to say, of course, that these essences are always worth investigating. Whether they are depends on how explanatory the kinds are. So the essence of being Australian is of little interest, that of biological taxa, of great interest, that of biological categories, of uncertain interest.

2.3 Three Important Distinctions

My argument for *Partly Intrinsic Taxon Essentialism* utilized three important distinctions which I will now identify. The discussion that follows will throw further light on these distinctions and their role in my argument.

1. The Distinction between Structural and Historical Explanations (2008, 351–5; 1.3 above): Why believe in *Partly Intrinsic*? The first part of my argument for it was that biological generalizations about the morphology, physiology, and behavior of a species or other taxon require explanations that advert to an intrinsic underlying, probably largely genetic, property that is part of the essence of the taxon. If we put together each such underlying property that similarly explains a generalization about a taxon, then we have the intrinsic part of its essence.

The explanations in question here are "structural" ones about the underlying states in members of a taxon that, along with the environment, cause members to develop in a way that *makes the generalizations true*. In urging this argument, I emphasize a crucial distinction, made by Ernst Mayr (1961) and renamed by Philip Kitcher (1984), between these structural explanations and "historical" ones about how members of the taxon *evolved to have* such states.[8] My argument is that *structural* explanations demand (partly) intrinsic essences:

> There has to be something about the very nature of the group—a group that appears to be a species or taxon of some other sort—that, given its environment, determines the truth of the generalization. (2008: 352; 1.3 above)[9]

And that something can't be a relational property. I was particularly harsh on the idea that the explanatory property might be that of standing in an appropriate relation to designated individuals, perhaps a "type specimen" (Hull 1978: 311–12)

[8] Mayr named the first sort of explanation "proximate", the second, "ultimate". Andre Ariew (2003), in a helpful critical discussion of the nature of this distinction, stays with "proximate" for the first sort of explanation but argues persuasively that "evolutionary" is appropriate for the latter. Still, I'm sticking with "historical".

[9] David Hull takes the standard argument against intrinsic essentialism to include the assumption: "The only basis for a natural classification is evolutionary theory" (1965: 203). My emphasis on structural explanations goes against this assumption.

or, "say, Brigham Young" (Ruse 1987: 344). I called this idea "explanatorily hopeless" (2008: 363; 1.7 above).

The second part of my argument is related to the first. Mohan Matthen points out that biologists think "that something is striped *because* it is a tiger" (1998: 115). Ereshefsky and Matthen ask, "Why does this bird have black feathers?", and answer, "Because it is a crow" (2005: 2–3). These remarks exemplify the just-made point that biological classifications are explanatory. I put the point like this: "the fact that an individual organism is a tiger, an Indian rhino, an ivy plant, or whatever, explains a whole lot about its morphology, physiology, and behavior" (2008: 352; 1.3 above). *Why does it*? Because for an organism to be, say, a tiger, is for it to have the intrinsic underlying nature of tigers and that nature, in combination with the environment, causes its phenotypic features, its morphology, physiology, and behavior: the same underlying properties that make the organism a tiger cause it to be striped. That's why being a tiger is explanatory.

In sum, the intrinsic nature of a taxon explains both the truth of generalizations about its members and why being in the taxon is explanatory.

An important methodological point: Leslie claims plausibly that the traditional argument for *Partly Intrinsic* "makes critical use of intuitions" (2013: 109). As can be seen, my argument does not. It makes critical use of biological explanations.[10] I should have emphasized this.

2. The Distinction between the Taxon and the Category Problems (2008: 356–63, 366–70; 1.5–1.7, 1.9 above): The argument from structural explanations is an argument *for Partly Intrinsic*. What does the consensus have *against Partly Intrinsic*? Okasha expresses the main objection: "On all modern species concepts (except the phenetic), the property in virtue of which a particular organism belongs to one species rather than another is a relational rather than an intrinsic property of the organism" (2002: 201). I argued that this objection is seriously mistaken, conflating another distinction due to Mayr, that between "the category problem" and "the taxon problem" (1982: 253–4). Mayr introduced the distinction for species, but we can generalize it. Then, the category problem is concerned with what it is for a taxon to be in any particular category, for example **Genus**. The taxon problem is concerned with what it is for an organism to be a member of a particular taxon, say *Canis*. The "species concepts" that Okasha is alluding to are theories that address the *category* problem *for* **Species**: they say what it is for a taxon to be a species (rather than, say, a subspecies or genus).[11] These concepts provide a *constraint* on an answer to the taxon problem *for any taxon that is a*

[10] So too does Walsh's argument (2006).

[11] Species concepts provide an "ontological definition of species". This is distinguished from epistemic "species delimitation", the way of *discovering* when a group of organisms form a species (Camargo and Sites 2013: 225).

species but fall far short of providing an answer. And the concepts *throw no light at all* on the problem for taxa that are not species. In contrast to these concepts, *Partly Intrinsic* is an answer to the *taxon* problem and has nothing to say about the category problem. So *Partly Intrinsic* and the concepts are answering different questions. The consensus, as expressed by Okasha, is simply wrong.[12]

Discussions of essentialism in biology are dominated by talk of species. This domination is unfortunate in two ways. First, it leads to the almost total neglect of taxon problems for non-species; so answers are not comprehensive. For example, taxon problems for non-species go totally undiscussed by most of my critics.[13] Second, the domination muddles the essentialism issue, which arises equally for taxa of all categories, with the vexed issue, addressed by the species concepts, of what it is for a taxon to be in the species category rather than in some other category. As noted, the essentialism issue concerns the property an organism must have to be a member of some taxon that has been identified and named by biologists and that is standardly thought to have a certain place in the Linnaean hierarchy. Whether or not that taxon has that place, whether or not it counts as a subspecies, species, genus, or whatever, is an issue that is quite independent of taxon essentialism.

How are the intrinsic essences of taxa in a hierarchy related? Where taxon *T* is in a lower category than taxon *T** and falls within *T**, the essence of *T* must include the essence of *T**, but not vice versa. The underlying essential property of *T** that explains why the members of *T** have phenotypic property *P* explains why the members of *T* have *P*; hence that underlying property is part of *T*'s essence. Thus, everything that has the essence of *Canis familiaris* has the essence of *Canis*, but not vice versa; some *Canis* are coyotes.

3. *The Distinction between the Conspecificity Problem and the Taxon Problems* (2008, 363–6; 1.8 above): *Why* has the significance of distinction 2 been standardly overlooked in discussing essentialism? My "tentative diagnosis" was that this significance has been overlooked because of the appealing, but mistaken, idea that the species concepts imply relational answers to the *conspecificity* problem, the problem of what makes two organisms conspecific. For, if that idea were right, then *Partly Intrinsic* could not be. But the idea is not right.

These distinctions 1 to 3, and associated arguments, are central to my response to my critics. These authors have made more criticisms than I can respond to in

[12] Mayr's distinction is well-established and yet, as I demonstrated (2008: 363–6; 1.8 above), its significance is standardly overlooked in discussing taxon essentialism; see section 2.6 below for more examples.

[13] Ereshefsky (2010: 674), Richards (2010: 66), Leslie (2013: 111), and Slater (2013: 48) mention these problems but never discuss them. Barker (2010) and Lewens (2012) do not even mention them. However, Wilson et al.'s (2007) "homeostatic property cluster" view and Godman et al.'s (2020) "historical essentialism" apply to all biological taxa.

detail here and so I have had to make some hard choices. I have chosen to make detailed responses only where those responses seem to me to advance the case for *Partly Intrinsic* rather than simply labor the case already made. In section 2.4, I respond to criticisms that *Partly Intrinsic* is at odds with certain biological variations. These criticisms come from Leslie (2013), Slater (2013), Wilson et al. (2007), Lewens (2012), and Richards (2010). In section 2.5, I respond to the criticism, first put to me by Peter Godfrey-Smith (as noted, 2008: 354; 1.3 above) and emphasized by Ereshefsky (2010), that *Partly Intrinsic*'s talk of intrinsic essences is an uncalled for metaphysical addition to biology. In section 2.6, I discuss Ereshefsky's (2010, 2014) handling of distinction 2 in his answer to the taxon problem for species. In sections 2.7 and 2.8, I discuss Barker's (2010) handling of distinction 3 in his answer to that same problem. His discussion supports the above diagnosis: I am no longer tentative about it. Finally, in sections 2.9 to 2.11, I discuss Godman, Mallozzi, and Papineau's (Godman and Papineau 2020; Godman et al. 2020) rejection of *Partly Intrinsic* in favor of the consensus view that species have historical essences.

In section 1.1 (point (v); 2008: 348), I noted that Michael Ghiselin (1974) and David Hull (1978) take their view that species are *individuals* and not kinds to be an antidote to essentialism. Ingo Brigandt claims that "most biologists and philosophers favor the idea that species are individuals rather than natural kinds" (2009: 77–8). Brigandt may be right about philosophers of biology—certainly the debate over type specimens and reference, discussed in Chapter 5, provides evidence that he is—but a recent survey (Pušić et al. 2017) shows he is quite wrong about biologists. The survey of the opinions of 193 biologists from over 150 biology departments at universities in the US and the EU found that, among biologists themselves, the position of individualism is "utterly marginal", only 2.94% (p. 197). In any case, I argued, in agreement with Okasha (2002: 193–4), that this individualism is a red herring to the essentialism issue.

The focus of this chapter is on defending *Partly Intrinsic* from criticisms. But part of the case for *Partly Intrinsic* is the lack of a viable relational alternative. I mentioned above some alternatives that I have argued are "explanatorily hopeless". What I should have emphasized more in "Resurrecting" was that the literature does not contain a *plausible worked-out* relational alternative. And, as we shall see, my critics do not provide one. The consensus is that a taxon's essence is to be found in its history but *where precisely* is it to be found? This question is not adequately addressed. Nor is the question of how any such relational essence could plausibly be thought to carry the explanatory burden. I develop this line of criticism in "Historical Biological Essentialism" (2018a) and in Chapter 3 below. I go on to argue that there is indeed an historical component to a taxon's essence but this component demands an intrinsic component. (This provides, of course, another argument for *Partly Intrinsic Taxon Essentialism*.)

2.4 Variation

2.4.1 The Common Cause Hypothesis

Leslie finds my argument from structural explanation to *Partly Intrinsic Taxon Essentialism* "intuitively appealing, but not... ultimately successful" (2013: 133). Focusing on variation, she makes a number of good points about underlying states and phenotypic features. She thinks that these points count against *Partly Intrinsic*. Indeed, intrinsic essentialism is standardly taken to overlook variation: "To dismiss variation as unimportant and to classify specimens into discrete categories is a manifestation of essentialism, a world view that Mayr (1963) has called 'typological thinking'" (Futuyma 1986: 108). The idea "that variation among organisms is the crucial stuff of changing life and of life's progress" is thought to be "devastating to essentialism" (Hey 2001: 62).[14] In contrast, I have argued that *Partly Intrinsic* accommodates variation nicely (2008: 370–8; 1.10 above). Leslie's variation objections are interestingly different from the ones I have discussed. Some related objections are made by Slater, Wilson et al., and Lewens. And Richards has another variation objection. I shall address these objections in turn. I think that *Partly Intrinsic* can handle them too.

Leslie considers my example of the rhinoceros and comments:

> Devitt is, of course, indisputably correct that *each particular African rhino* has some intrinsic features that, in combination with the environment, are causally responsible for that individual's having horns. This does not entail, however, that those very same intrinsic features are also responsible for other African rhinos' having horns. Whether this is so is a substantive empirical hypothesis, not one whose truth can be intuited in advance. (2013: 134)

Leslie supports this with the case of jade. As is well known, two different chemical compounds, jadeite and nepthrite, have been lumped together as jade. So, no common intrinsic chemical structure explains the similar observable features of all samples of jade. Jade has a "disjunctive" essence.

Leslie is right, of course, that the fact that some intrinsic feature causes one rhino to have horns does not *entail* that that feature causes other rhinos to have horns. But the argument for *Partly Intrinsic* does not rest on any such entailment. It rests on what, *as a matter of fact, explains* the (nonaccidental) fact that rhinos have horns. The claim that some one intrinsic feature of rhinos is responsible for this is, as Leslie says, "a substantive empirical hypothesis". But the modal force of

[14] Cf. "as a general principle essentialism is wholly compatible with substantial change, a phenomenon continuously exhibited in the inorganic world. It is an elementary mistake to think that fixed essences exclude substantial change" (Oderberg 2007: 204).

the hypothesis arises from an "inference to the best explanation" not an entailment. The argument for *Partly Intrinsic* rests on fallible empirical hypotheses of just this sort. I claimed that such hypotheses, implicitly embraced even by little children (Keil 1989; see also Gelman 2003), are implicit in the practice of taxonomists (2008: 352–3). In particular, I now claim, these hypotheses are supported by the contemporary role of genetic analysis in *re*classification.[15]

Take tigers (*Panthera tigris*), for example. A paper begins with the following claim about what were thought to be the several subspecies of tiger: "available molecular evidence suggests that extant tigers are extremely similar genetically" (Cracraft et al. 1998: 139). Nonetheless, using DNA sequencing techniques, the paper argues, *on the basis of genetic differences*, for a reclassification: the Sumatran tiger is not a subspecies of *Panthera tigris* but a distinct species (1998: 148).

Similarly, a paper on African elephants argues against "the consensus that all belong to the single species *Loxodonta africana*" largely on the basis of a "deep genetic division between the forest and savannah populations" (Roca et al. 2001: 1473–4). The paper concludes that these populations form two species, the former, *Loxodonta cyclotis*, and the latter, *Loxodonta africana* (2001: 1476).

"Cryptic species" provide many examples of taxonomic reclassification on the discovery that phenotypic properties that were thought to be caused by the one underlying nature are actually caused by two:

> Cryptic species are defined as "two or more distinct but morphologically similar species that were classified as a single species" (Pfenninger and Schwenk 2007). Because of this morphological similarity, most cryptic species were initially considered to be a single species until genetic data and rigorous scrutiny of phenotypic characters indicated otherwise. Cryptic species are relatively common across a wide range of taxa and habitats. (Andrews et al. 2016: 361)

What we see is a pattern of tying a taxon to an underlying genetic structure, a structure that causes its phenotypic properties, and of reclassifying whenever it is discovered that there is significant genetic difference among two populations within a taxon. And this is just what we should expect because the underlying structure makes the taxon explanatory (2.2, 2.3).

Slater thinks that taxonomic errors of the sort just illustrated pose a problem for my essentialism (2013: 48), but they do not. Thus, in cases like the Sumatran tiger and the forest elephant, the one taxon with a certain underlying essence is simply moved from one Linnaean category or rank (**Subspecies**) to another (**Species**). In the case of cryptic species, a group that was once thought to be a taxon in a Linnaean category (**Species**) has been determined not to be. Since it is not, *Partly*

[15] Thanks to Derek Skillings for drawing my attention to some of this literature.

Intrinsic does not apply to it. This is not to say that the group lacks an essence: like jade, it has a disjunctive essence; see subsection 2.4.6 below.

It is indeed "a substantive empirical hypothesis" that members of what is thought to be a Linnaean taxon share an intrinsic underlying property that causes its phenotypic properties. Yet it is a *prima facie* plausible hypothesis. Consider any particular organism in that taxon. What explains the fact that it developed with the characteristic phenotypic properties of the taxon? Some intrinsic underlying property at the genetic level together with the environment. That is surely undeniable. Then, the plausible essentialist hypothesis is that the *very same* underlying property, whatever it may be, explains those phenotypic properties in other members of its taxon in that environment.

Indeed, what plausible alternative is there to the idea that there is a common underlying cause of the phenotypic properties used to identify a taxon? Surely not that there is a large variety of underlying causes. We might, of course, discover that more than one underlying cause is operative, as cryptic species illustrate. So, our hypothesis might be wrong, like any empirical hypothesis. But that does not gainsay its plausibility. And such hypotheses have been confirmed in many cases by genetic analysis.

2.4.2 Genetic Variations

Leslie claims that "a member of one species . . . may have more genetically in common with a member of another species . . . than with a member of its own species" (2013: 133). The argument for *Partly Intrinsic* requires that members of a species share a genetic structure that largely constitutes the nature of the species. It does not require that the rest of the genetic structures of these members be the same. So *Partly Intrinsic* is quite compatible with what Leslie claims. What matters to *Partly Intrinsic* is not an overall similarity among taxon members, established by "counting genes", but a similarity in one particular part of the underlying structure, the part that causes (in its "normal" environment) the distinctive phenotypic features of the species. (I make a similar response to a similar objection to racial realism; see section 6.5.2.)

In a passage that I discussed in "Resurrecting" (2008: 370–2; 1.10 above), Wilson also emphasizes genetic variation within a species. He claims that essentialism cannot accommodate the "inherent biological variability or *heterogeneity* of species" that is "a cornerstone of the idea of evolution by natural selection" (1999b: 190). He has since returned to this criticism of what he calls "traditional essentialism", joined by Barker and Brigandt, and citing Devitt (2008) as an example (Wilson et al. 2007: 189). The "fundamental reason" why this essentialism is mistaken is that "biological kinds", unlike "those of the physical sciences" are

> *intrinsically heterogeneous* in that the individuals they subsume do not simply differ from one another in the properties they possess, but do so *by nature or intrinsically as things of that kind*.... this variation...plays an important causal role in many biological processes, not the least of which is natural selection itself. (p. 192)

But this fundamental reason is fundamentally wrong: the essence of a biological kind *explains* the intrinsic heterogeneity that is "crucial to the underlying causal mechanisms at the heart of biological stasis and change" (p. 193). For, biological variation in individuals is a consequence of the mechanisms of reproduction, and the underlying essence of the kind (along with the environment) is the cause of those mechanisms. The essences of biological kinds are thus central to the causal story of natural selection, not at odds with it. There is no such story for other kinds because they have different sorts of essences. (Ironically, this defense of *Partly Intrinsic* is analogous to a defense that Wilson et al. make (p. 210) of their own homeostatic property cluster essentialism.)

2.4.3 Phenotypic Variations

Leslie's next objection moves from genetic variation to phenotypic variation: "conspecificity is compatible with a great deal of variation in phenotype at a time, and even more dramatically over time" (2013: 134). The argument for *Partly Intrinsic* hypothesizes an underlying nature that causes the phenotypic properties that are *shared* among conspecifics. Conspecifics can of course share those phenotypic properties *while differing in many others*. Thus, sexual dimorphism is common; and there are stable developmental and seasonal polymorphisms.[16] Phenotypic difference within a taxon is even more obvious when we think of the higher ones (remember, *Partly Intrinsic* applies to them too); thus, the primate taxon (an **Order**, in Linnaean ranks) includes lemurs as well as humans, which differ greatly in phenotype. The phenotype variations noted by Leslie do not undermine *Partly Intrinsic*.

Those variations are caused by intrinsic underlying differences within a taxon. But, of course, phenotypic variations can also be caused by environmental differences. Consider the standard example of air temperature determining the sex of fetuses in some turtles and reptiles; the impact of temperature, soil, etc. on the properties of plants is obvious; hemlocks growing in the mountains are short, those at lower altitudes, tall (Ereshefsky and Matthen 2005: 5). From the

[16] Ereshefsky and Matthen charge that the homeostatic property cluster theory is unable to explain these polymorphisms (2005: 7–10). Wilson et al. (2007: 210–11) respond effectively to this charge, in my view. I would respond similarly to any such charge aimed at *Partly Intrinsic*.

perspective of *Partly Intrinsic*, there is no puzzle about this. The one underlying nature of a taxon causes its members to have certain features in one environment, others in another.

This prompts two important points. First, the underlying nature of a taxon is just as explanatory of the *atypical* phenotypic features of a member produced by an "abnormal" environment as of the *typical* features of members in their "normal" environment. The nature plays its causal role across *all* environments. (And, probably, no phenotypic feature is essential.) Second, my main argument for *Partly Intrinsic* was that the intrinsic essence of a taxon explained generalizations about the taxon, for example, that Indian rhinos have one horn (2008: 351–5; 1.3 above). Clearly these generalizations are implicitly restricted to the taxon's "normal" environment.[17]

Slater sees the causal role of the environment as generating a problem in taking "genetic structure" as the essence of a species just as we take chemical structure to be the essence of a chemical kind:

> there is an important contrast between . . . chemical structure and an organism's "genetic structure" . . . . An organism's "genetic structure" . . . does not alone determine, or even strictly suffice to explain, facts about the organism's outward character or behavior in the absence of information about its actual environment. (2013: 44)

But there is no significant contrast. Change the environment of a chemical kind enough and its outward character or behavior may change; thus, zinc ceases to be malleable below 300 degrees Fahrenheit.

2.4.4 Causes of Phenotypic Properties

Leslie points to variation in the causes of phenotypic features: "In general, within a given species, individuals who share a common phenotypic feature need not share intrinsic microstructural bases that gave rise to the feature" (2013: 136). A feature that is "genetically induced" in some individuals might be "environmentally induced" in others (2013: 137). And so it might.

[17] I noted this (2008: 377, n. 50; 1.10, n. 58 above) but did not emphasize the related point that "abnormal" environments cause exceptions to the generalizations. (The implicit restriction does not imply any commitment to the Aristotelian "Natural State Model" (2008: 371–2; 1.10 above).) Mutations can also cause apparent exceptions to these generalizations, of course: "monsters". Leslie seems to think that *Partly Intrinsic* cannot allow for these exceptions and hence is "highly revisionary" (2013: 140, n. 23). I argued that this is not so (2008: 375–6; 1.10 above).

(a) Consider environmentally induced phenotypic features first. We have just given some examples. Leslie provides a nice further one, the Himalayan rabbit, a breed of the Common Rabbit (*Oryctolagus cuniculus*):

> Himalayan rabbits, when raised in moderate temperatures, have white body fur with black tails, noses, and ears; if they are raised in cold temperatures, however, they develop wholly black fur. (2013: 137)

From the perspective of *Partly Intrinsic*, there is no puzzle about this. The one underlying nature of this breed causes its members to have certain features in moderate temperatures and certain other features in cold temperatures.

(b) Now consider genetically induced phenotypic features. Leslie points out that the black fur of a rabbit can arise not only from the environment, as with the Himalayan rabbit, but "relatively straightforwardly from a given rabbit's genetic make-up—that is, as an inherited trait that manifests itself across various environments" (2013: 137). But note that such a rabbit's black fur does *not* arise from the part of its genetic make-up that is essential to its being a rabbit (else rabbits would typically be black). So, the underlying nature of the taxon *rabbit* does not determine that fur color. So this phenomenon has no bearing on *Partly Intrinsic*.

In sum, the underlying nature of a taxon causes different phenotypic properties in different environments. Different genetic make-ups among organisms that share the common underlying nature of a taxon can lead to different phenotypic properties. Indeed, the latter point is obvious when we remember that both lemurs and humans are in the primate taxon.[18]

2.4.5 Complicated Developmental Pathways

Leslie's most interesting objection to *Partly Intrinsic* is to be found in her vivid discussion of cases in support of the following:

> Phenotypic traits are the upshot of complex biochemical processes controlled in most cases by a variety of genes. Differences in the genetic level need not translate into differences in the biochemical processes... Canalization of a trait insures that the trait is stably expressed in the face of underlying genetic variation... we might say that phenotypic traits often exploit a certain *multiple realizability* at the

[18] John Wilkins thinks that the notion of an essence consisting of a "set of shared properties... common to every member of a taxon... is biologically unnecessary" (Wilkins 2010: 146). The common underlying nature is necessary to explain the phenotypic properties of members of the taxon in an environment.

> microstructural level.... Every macroscopic phenotypic property depends on a massive number of biochemical reactions, originating with the genes themselves but continuing along the entire developmental pathway, at each point potentially subject to environmental influences, influences from other genes, and so on. (2013: 138)

We have just discussed the impact of the environment on the developmental pathway to a phenotypic property in species members and need say no more. But what about the impact of the genes?

Leslie is emphasizing that the path from genes to a particular phenotypic property is both complicated and varied. Let us start with the complication: "a variety of genes" is involved. *But it is no part of the argument for Partly Intrinsic that the paths from genes to phenotypic traits are simple.* Indeed, although for convenience I sometimes talk simply of a genetic essence, I am noncommittal on precisely what underlying properties constitute the essence of a taxon and hence cause its phenotypic properties:

> In sexual organisms the intrinsic underlying properties in question are to be found among the properties of zygotes; in asexual ones, among those of propagules and the like. For most organisms the essential intrinsic properties are probably largely, although not entirely, genetic. Sometimes those properties may not be genetic at all but in "the architecture of chromosomes," "developmental programs," or whatever (Kitcher 1984: 123). (2008: 347; 1.1 above)

Indeed, it would be foolhardy for *Partly Intrinsic* to have commitments on these empirical biological matters, and it has none. We already know enough, of course, to reject "the crude idea that there is, say, 'a tiger gene'" (2008: 371; 1.10 above). Okasha thinks "that species are distinguished by clusters of covarying [chromosomal and genetic] properties" (2002: 197). So perhaps, I suggest (2008: 371; 1.10 above), that is where we should look for the underlying essence. But perhaps not. Consider these interesting claims by Denis Walsh in the course of arguing for intrinsic essences:

> We have grown accustomed to thinking of genes working together as pathways. The development of each character can be traced back through a fairly discrete causal pathway to the actions of a small set of genes. This view of gene function, however, has recently been shown to be quite untenable, at least for an extremely large part of the organism's genome. Rather, genes are organised, not so much as pathways, but as complex, regulatory networks. Phenotypes are produced through the complex interactions among a number of gene-regulatory networks (inter alia). No phenotypic feature can be traced to any particular gene or subset of genes within the network. (2006: 436)

The moral of this is that the wise philosophical essentialist leaves the details of essences to scientists; biological essences are to be discovered by biologists.

Aside. This point generalizes. Leslie criticizes Kripke-Putnam essentialism not only about *biological* kinds but also about *chemical* kinds (2013: 142–58). Her criticism draws heavily on Paul Needham's long-running campaign against Kripke-Putnam essentialism about water; see Needham (2011) and the earlier papers cited. This campaign, based on an illuminating presentation of the scientific facts about water, includes criticism of the Kripke-Putnam talk of "H_2O" as the essence of water.[19] But it is central to the positions of Kripke (1980: 119–29) and Putnam (1975: 224–5) that we should look to *empirical science*, not to philosophers like themselves to fulfill the supremely important task of *discovering* the essence of "natural kinds" like water. Needham is persuasive that the H_2O-talk is inadequate, but a recent paper by another philosopher of chemistry, Robin Hendry (forthcoming), suggests that it may not be. In any case, the H_2O-talk should be seen as nothing more than a philosopher's hand wave toward the scientific facts.[20]

I conclude that a complicated role for genes in causing phenotypic traits is not a problem for *Partly Intrinsic*.

2.4.6 Disjunctive Developmental Pathways

Leslie is also emphasizing something that may seem more problematic for *Partly Intrinsic*: the genetic component involved in the developmental pathway to a phenotypic property can vary among the members of a species. Thus, commenting on the cause of human female genitalia, Leslie says:

> For each individual infant, there will be a genetic component to the explanation (as well as an environmental component), but this genetic component need not be the same for each infant. (2013: 136)

[19] Talk of water being H_2O was, of course, quite standard in philosophy at that time; see discussions of "the identity theory" of mind, for example.

[20] Needham is also critical of the Kripke-Putnam empirical conjecture that the essence of water is an underlying, *entirely micro*, matter. Needham does not convince me that this conjecture is wrong, but even if it is and the essence is partly, even entirely, macro, that alone would not be much of a blow to the metaphysics of Kripke and Putnam. They surely already accept that many scientific kinds have macro essences; *predator* is a likely example. What *would* be a blow would be the discovery that the essence of water is constituted entirely by the macro properties *picked out by descriptions associated with the word 'water' by its ordinary users*. For, *that* is the consequence of the description theory of reference that Kripke and Putnam are mainly challenging: in the clever words of Putnam's slogan, the meaning of 'water' "just ain't in the head". Needham does not attempt to undermine that challenge. For more on this see Devitt (2021c: 421–3).

She makes similar remarks about the three toes on the hind feet of guinea pigs:

> *having three toes on the hind feet* is a characteristic property of guinea pigs (*Cavia porcellus*). Possession of this phenotypic property is due to a flexible interaction-effect between a number of factors, both genetic and non-genetic—that is, there is a generous range of pathways, all of which lead to having three hind toes. Individual guinea pigs can differ significantly from each other with respect to these factors and yet each have three toes... (p. 137)

See also her discussion of the number of eggs produced by female salamanders (pp. 137–8). The previous untroubling point was that many genes are involved in causing a phenotyptic property. The present point is that the ones involved can vary within a taxon: the posited essence of a taxon can cause a phenotypic property in more than one way. Walsh is illuminating on this. He continues the above passage on regulatory gene networks as follows:

> These networks, or circuits, are characterised by flexibility, redundancy and robustness (Salazar-Ciudad et al. (2001); Szathmary (2002)). All of these features function to preserve the capacity of a gene network to produce its normal output in the face of perturbations and variations of initial conditions. For example, gene regulatory networks show a remarkable capacity to compensate for the 'knock-out' of elements of the network. If a gene is removed, or 'knocked-out', of a regulatory network the network typically compensates and finds alternative ways of producing its characteristic output (Greenspan (2001)). (2006: 436–7)

Walsh does not see this variation in the effective genetic component as a problem for intrinsic essentialism. Quite the contrary. He takes the essence of a species to be "the causal capacities of its developmental systems" which realize its "phenotypic plasticity" (2006: 441). And that plasticity consists in just the sort of "flexibility, redundancy and robustness" illustrated by Leslie's guinea pig. So, why does Leslie think that the variation is a problem?[21]

Leslie objects to the "disjunctive" nature of the underlying cause of a phenotypic property, its being *either* this genetic component *or* that one. Using the example of jade again, Leslie argues that *Partly Intrinsic* cannot allow this:

> the observable properties of the sample of jadeite are determined by its intrinsic chemical structure in conjunction with the environment, and similarly for the sample of nephrite. Yet there will be no common intrinsic chemical structure that explains the shared features of the two samples of jade.

[21] Leslie does not mention Walsh's paper.

> (There will, of course, be the non-explanatory disjunctive property of being composed of *either* NaAl(SiO3)2 *or* Ca2(MgFe)5Si8O22(OH)2. However, it is important to see that disjunctive properties cannot play the explanatory role that Devitt has in mind, or else the whole enterprise is trivialized. For example, let us suppose with Devitt that there is a common intrinsic property had by tigers that explains why they are striped. Let us also suppose that there is a different common property that explains why canna lilies are striped. If disjunctive properties are allowed to figure as common intrinsic explanatory properties in Devitt's sense, then there will be a further shared intrinsic property that explains why *this tiger* and *this lily* both have stripes. If disjunctive properties are countenanced in this endeavor, then *shared* properties become far too cheap to be of interest. Certainly, it would not then be a *biological* hypothesis that a common property explains why Indian rhinos have one horn—it would simply be a familiar point about the logic of disjunction.) (2013: 134)

Lewens makes a similar point, claiming that there might be various microstructural causes of stripyness in tigers and continuing:

> Of course, we will be able to identify a characteristic 'pattern' of genotypic properties across the species, simply by enumerating whichever genes actually cause the instances of species-typical phenotypic properties, but since this must be the case however unruly these underlying genes are, the sense of essence we salvage here is trivial. (2012: 753)

This is interesting.[22] But the wrong moral to draw from it is that the one underlying nature cannot yield disjunctive explanations of observable properties. We need to distinguish such disjunctive explanations, which arise from *a non-disjunctive "unified" essence*, from ones arising from *a disjunctive essence*. Jade offers explanations of the latter sort. Why do these two samples of jade have a certain observable property? The explanation for one sample is that it has a certain underlying property, the essence of jadeite; the explanation for the other sample is that it has a quite different underlying property, the essence of nephrite. But, as Leslie points out, there is "no common intrinsic chemical structure that explains the shared features of the two samples of jade" (2013: 134). And that is what makes jade a scientifically unacceptable kind: it provides explanations from a disjunctive essence: instances of jade have *either* the essence of jadeite *or* that of nephrite, *but not both*. So, alluding to jade does not provide scientifically adequate explanations; it does not "carve nature at its joints".

[22] The following response is, I hope, an improvement on the one in Devitt (2021a: 13–14).

Turning to biology, for just the same reason, alluding to the kind tiger-or-canna-lily does not provide scientifically adequate explanations. Less fancifully, alluding to a cryptic species does not. A cryptic species provides explanations from a disjunctive essence. And that is why a cryptic species is not a biologically acceptable species.[23] But none of this shows that biology does not tolerate disjunction in the way that the one underlying nature causes a taxon's phenotypic properties. Indeed, the moral of Leslie's cases of female genitalia, the guinea pig, and the salamander is precisely that biology embraces disjunctive explanations: the underlying nature provides several different causal paths to a phenotypic property. This causal flexibility is part of the essence, as Walsh emphasizes.

So, what is the difference between the acceptable disjunctive explanations of a phenotypic property provided by the guinea pig and the unacceptable ones provided by a cryptic species? There is one underlying property common to guinea pigs, the complicated essence of guinea pigs, that can cause the phenotypic property in more than one way. There is not one such underlying property common to members of a cryptic species (or tiger-or-canna-lily). For, the cryptic species is a kind made up of two explanatory sub-kinds, two species. So, there is one underlying property common to the members of one of those species, its complicated essence *E1*, that causes the phenotypic property in its members, a different underlying property common to the members of the other species, its complicated essence *E2*, that causes the phenotypic property in its members. One might be tempted to say that the disjunctive essence of the cryptic species, *E1-or-E2*, explains the phenotypic property in all its members from either species, just as one might be tempted to say that the disjunctive essence of jade explains observable properties of jade, whether jadeite or nephrite. But one should not give in to these temptations because these are not *good* scientific explanations.

Some counterfactuals may help. Any particular tiger's property of being striped is caused *in a certain way* (in its environment) by the underlying properties constituting the essence of tigers; similarly, any particular canna lily's, *in a certain different way*, by the essence of canna lilies. It is not the case that the stripyness of that tiger might have been caused in the different way of the lily's, nor that the lily's stripyness might have been caused in the different way of the tiger's. In contrast, any particular guinea pig's property of having three toes on its hind feet is caused *in a certain way* (in its environment) by the underlying properties constituting the essence of guinea pigs; similarly, another guinea pig's, *in a certain different way* (let's suppose), *also* by the essence of guinea pigs. Yet either guinea

[23] With cryptic species in mind, Slater objects to *Partly Intrinsic*: "it is not clear to me that disjunctive essences would play the robust explanatory role Devitt reckons essences must play" (2013: 48). But *Partly Intrinsic* requires that "robust" role of the essences of species not of groups that were wrongly thought to be species.

pig's property of having the three toes might have been caused in the different way it was caused in the other guinea pig.

In sum, the argument for an intrinsic essence presented in section 2.3 holds: it is *because* of the underlying largely genetic essence of guinea pigs that they have three toes on their hind feet; and it is *because* a certain animal is a guinea pig, hence has the underlying nature of a guinea pig, that it has three toes on its hind feet. These are biologically sound explanations. But similar essentialist explanations of a phenotypic property of a cryptic species are not acceptable. And that's why cryptic species are not species.

I conclude that the disjunctive explanations that Leslie and Lewens are pointing to do not undermine *Partly Intrinsic* but rather throw an important light on a species' intrinsic explanatory essence. The explanations demonstrate that not only is the essence of tigers not "a tiger gene", it is a long way from a tiger gene.

2.4.7 Evolving not Timeless

Finally, I turn to Richards' variation objection. He thinks that an essentialism like mine faces the following "common criticism":

> Natural kinds are (presumably) timeless and eternal, and the set of essential properties that make an organism a natural kind is also timeless and eternal. But evolutionary change implies a change in the properties of organisms. Hence, species cannot evolve if they are natural kinds. (2010: 156)

What to make of this talk of "natural kinds" that are "timeless and eternal"? I am wary of talk of "natural kinds" (though I indulge occasionally; e.g. 2011c) because I think it is rather unclear what "natural" means here. So let us consider kinds in general. On my view, kinds, whether "natural" or not, have essences; for example, there is an essence to being gold, a pencil, and Australian (1.1; 2.9). Are these essences "timeless and eternal"? This seems an odd question, but I suppose that the answer is "Yes": *what it is to be* a pencil does not change from week to week. What does change over time, of course, are *the objects that are* pencils. Even the *sorts* of objects that qualify as pencils, the *sub*-classes of pencils, may change; thus, there were once no propelling pencils. And just the same is true of taxon essences: being a tiger is "timeless and eternal", in this uninteresting way, but tigers, and subspecies of tigers, may come and go (several subspecies are now extinct; Cracraft et al. 1998: 139). So, no sign yet of any problem for *Partly Intrinsic* with evolution.

Perhaps anticipating this response, Richards continues:

> Devitt has a response ready, though, arguing that even though the natural kind cannot itself evolve, given that its essential properties are timeless and

> unchanging, organisms within a lineage can still change – by passing from one kind to another. (2010: 156)

Richards quotes a passage in which I talk of species *S1* with essence *G1* gradually evolving into *S2* with *G2* (2008: 372; 1.10 above). He notes that I deny "any sharp line between the essences *G1* and *G2*" (2010: 157), quoting a lengthy passage where I emphasize that organisms between *S1* and *S2* will not determinately have *G1* or *G2* (2008: 373; 1.10 above).

> This response by Devitt demands an answer to the question how the essences *G1* and *G2* are determined – if they are distinct from the group of organisms in the way suggested. If groups of organisms vary gradually, and we determine essences from observation of these organisms, as he suggests, then we have to decide which time is determinative. In gradual change, there will be a difference in properties from one time to another. If we don't treat some particular time slice as determinative, then the essences will change as the individual members of the species change. (2010: 157)

"Determine" is notoriously ambiguous between the epistemological "how we tell" and the metaphysical "what constitutes". Both senses seem present in Richards' passage. The sense in my talk of "indeterminacy" is the constitutive one and I'll start my response with that sense.

In "Resurrecting" (2008: 376; 1.10 above), I included a sentence from Hull that demonstrates how indeterminacy is built into evolution: "Obviously ... there must have been instances in which non-horses (or borderline horses) gave rise to horses" (1978: 306). Thus, it should be uncontroversial that before and after there were members of *S1* there were organisms in the *S1–S2* lineage that were not determinately members or non-members of *S1*. *This is so whatever one's view of essences.* Earlier Richards has claimed that, despite the "initial plausibility" of my essentialism, it is at odds with "what systematists actually do in grouping organisms into species. If a tiger mating pair has an offspring that lacks stripes, systematists do not therefore conclude it is not a tiger" (2010: 153). Similarly, a "rhino" that "does not in fact have a horn, perhaps because of a mutation in a developmental gene network ... is still an Indian rhino" (p. 154). The indeterminacy that accompanies evolution shows that "what systematists actually do" is not quite so simple.

Richards wonders "how the essences *G1* and *G2* are determined", which "particular time slice" is "determinative". And the answer for *G1*, which is the essence of *S1*, is that *G1* is a property that is common and peculiar to the organisms in the time slice of the *S1–S2* lineage that includes all and only *determinate* members of *S1*. Which property? The one playing the fundamental causal role described. Since the boundaries of this time slice are indeterminate—

see Hull above—so too are all the boundaries: those between *G1* and the indeterminates, between the indeterminates and *G2*, and between *G1* and *G2* (as I noted: 2008: 373; 1.10 above). It is that somewhat indeterminate *S1* time slice that is "determinative" of *G1*. And the essence of *S1* does *not* "change as the individual members of the species change". So, there is no problem about a taxon changing essences. Rather there is an indeterminacy about the essence of a taxon *that everyone must accept*, given evolution. Evolution leads to taxa having indeterminate essences not changing ones.

It is important to note that the indeterminacy arising from gradual evolutionary change can be matched by indeterminacy "in space", as John Dupré, in effect, demonstrates with ring species:

> The herring gull (*Larus argentatus argenteus*) and the lesser black-backed gull (*L. fuscus*) are two very familiar and quite distinct species, not known to interbreed. Yet it appears that if we track round the globe at roughly the same latitude, there exists a series of gradually diverging species each member of which is capable of interbreeding with the next but at the end of which are the herring gull and the lesser black-backed gulls. (2017: 231)

Just as there are organisms that are not determinately members of *S1* or its successor *S2*, there are organisms that are not determinately members of one of these ring species or its neighbor.

That is one conclusion to draw about ring species. But Dupré, in his campaign against essentialism, draws a different one: "In biology, it appears, distinct kinds are not given to us by nature but rather by our local and limited perspective on nature" (p. 231). In "Resurrecting" (2008: 374–5; 1.10 above), I remarked also that with gradual evolutionary change there can be some *arbitrariness* about which groups to choose and name as species. The truth underlying Dupré's conclusion is that there can be a similar arbitrariness about which groups among ring species to choose and name as a species. But, as before, the arbitrariness should not be exaggerated. Distinct kinds, albeit with indeterminate boundaries, *are* "given to us by nature" to the considerable extent that it is in virtue of having the essences of those kinds that organisms play their causal roles. And the key point is that such arbitrariness as remains "poses no threat to *Essentialism*" (2008: 374).

The significance of indeterminacy to biological essentialism is often misjudged. Thus, Dennett's (2017) recent tirade against this essentialism rests on the quite mistaken idea that essentialism is committed to "sharp boundaries". As Dennett notes, essentialism supposes that there is some property "in virtue of which [all *F*s] are *F*s" (p. 9). But to suppose this is *not* to suppose that it is always a determinate matter of fact whether or not an organism has that property; it is *not* to deny that "sets of living things" have "fuzzy boundaries" (p. 10); it is *not* to claim that "every animal either is a dog or isn't a dog" (p. 12). As Sober points out, "Essentialism is in principle consistent with vague essences" (Sober 1980: 253).

So much for metaphysical determination. Turn now to epistemological determination. How do "we determine essences from observation of these organisms"? Well, biologists examine organisms in the somewhat indeterminate time (even space) slices that mark out taxa to discover the properties playing the fundamental causal roles described.

Richards sums up:

> The problem is that these essentialists are asking us to think about things that have a history and change over time, in terms of a metaphysics that does not obviously reflect change. If our metaphysics tells us that the basic, fundamental things are unchanging, but our science tells us that there is change, and that change is a fundamental part of nature, there is a discordance that counts against *either* the metaphysics or the science. (2010: 158)

A kind, whether *pencils* or *horses*, has a "timeless" essence but it also has a history of change over time in that members come and go and they differ (think of the kind *NY Yankees*, for example). So this alleged problem for *Partly Intrinsic* seems to be a confusion (perhaps one between the members of a kind and the property of being a member).

I conclude that *Partly Intrinsic Taxon Essentialism* is not at odds with the biological variations identified by my critics. Thus, according to *Partly Intrinsic*, the underlying nature of a taxon causes different phenotypic properties in different environments; different genetic make-ups among organisms that share a taxon's underlying nature can lead to different phenotypic properties; the many genes involved in causing a phenotyptic property can vary within a taxon.

2.5 "The Added Metaphysical Claim"

Marc Ereshefsky is quite unconvinced by my argument from structural explanations for *Partly Intrinsic*. He concludes his criticism of it with a ringing reaffirmation of the consensus:

> To put it more starkly: the occurrence of certain relations *is* the species. It is the occurrence of those relations that makes the organisms engaged in them members of one species versus another species. In sum, intrinsic properties are not part of a taxon's membership conditions, and relational essentialists are right about how biologists individuate taxa. (2010: 683)

So, what does he have against my argument?

The structural explanations that concern this argument (repeated in section 1.3 above) are of generalizations like one that Ereshefsky considers, the generalization that zebras have stripes. Ereshefsky starts his discussion of this example as follows:

> embryonic zebras have developmental mechanisms that cause zebras to have stripes. These mechanisms are intrinsic features of embryonic zebras. But those developmental mechanisms must be passed down from parent to offspring via genealogical relations. So a robust explanation of why zebras have stripes cites both the relations and intrinsic properties that cause stripes. Merely citing relations provides a relatively weak explanation of an organism's trait.
>
> (2010: 680)

This is Ereshefsky's answer to what he calls "the trait question". In effect, he is pointing out that the *structural* explanation of why zebras have stripes adverts to intrinsic properties whilst insisting that the *historical* explanation must advert to genealogical relations.[24] So far, we are in agreement. But Ereshefsky plays down the significance of this explanatory difference with his talk of "robust" and "relatively weak" explanations. The structural and historical explanations *have different explananda*: they *both* offer robust explanations, *but of different things*, as I emphasized in making distinction 1 in section 2.3 above (see also, 2008: 351–5; 1.3 above). In my view, citing relations does not provide a *relatively weak* explanation of what it is about zebras that causes them to develop stripes; it provides *no* explanation at all. The explanation needed for that is a structural one adverting to intrinsic properties of zebras. I shall return to this point at the end of section 2.6.

Ereshefsky's major disagreement is with the move I make from my structural answer to his trait question, to intrinsic essences and hence to *Partly Intrinsic*. Applying this move to Eresheksy's example, I would claim that the intrinsic features of the embryonic zebras are partly constitutive of the essence of zebras. And it is because such intrinsic features are part of the very nature of zebras that the property of being a zebra is explanatory of those animals having stripes. Any animal that is a zebra has the essential properties of a zebra and these properties cause it (in its "normal" environment) to have the phenotypic properties that we expect zebras to have (2008: 352–3; 1.3 above).

Here is Ereshefsky's rejection of this move to *Partly Intrinsic*:

[24] Initially, this also seems to be Barker's view of structural and historical explanations (2010: 86). Yet, on the basis of appeals to species concepts, he claims that I must "acknowledge that evolutionary biologists propose or imply widely accepted wholly relational explanations of what *makes* structural generalizations true (rather than just explanations of what *led* to their being true)" (2010: 89). I don't acknowledge this because I don't think that it is so. And species concepts are beside the point; see section 2.6 below.

> Given the observation that we should cite both genealogy and developmental mechanisms to understand why zebras have stripes, should we infer, as Devitt does, that the taxon Zebra has an intrinsic essence? I do not think so. Biologists explain the characters of organisms by citing other characters, without the added metaphysical claim that the character cited in the explanans is essential to membership in a taxon. (2010: 680)

This disagreement is reminiscent of a nice question put to me by Peter Godfrey-Smith when I was writing "Resurrecting". I expressed the question like this:

> It is of course the case that the truth of any... generalization [about the phenotypic properties of a taxon] must be explained by an intrinsic, probably largely genetic, property, but why does that property have to be an *essential* property of the kind in question? (2008: 354; 1.3 above)

I attempted to answer this question in "Resurrecting" by appealing to the law-like nature of the generalizations and subjunctive conditionals (2008: 377–8; 1.10 above). Both Lewens (2012: 755–6) and Slater (2013: 51–3) do a good job of showing that this answer is inadequate. So I abandoned it (2021a: 14).[25]

Implicit in my argument for *Partly Intrinsic* was another answer that I strangely overlooked. A clue to this answer was given in subsection 2.4.3: "The nature plays its causal role across *all* environments." Any organism in taxon T has certain phenotypic properties *because* it is in T; it is *because* an animal is a zebra that it is striped. So the property that *makes* something a zebra, whether we call that property an "essence", "nature", or whatever, must *cause* that zebra, in its environment, to have stripes; the essence of zebras must explain the place that certain organisms have in the causal nexus *just because they are zebras*. Elliott Sober is getting at this with the following demand, which I shall call "the Sober demand": an essence "must be *explanatory*... A species essence will be a causal mechanism that acts on each member of the species, making it the kind of thing that it is"

[25] Ereshefsky has a puzzling *particular* reason for resisting the added metaphysical claim:

> In its embryonic state, a zebra has an ontogenetic mechanism that causes it to develop stripes. That developmental mechanism is neither necessary nor sufficient for membership in Zebra. (2010: 680)

What is the point of the claim that the mechanism is not *sufficient*? *Of course*, "the developmental mechanism that causes stripes in zebras causes stripes in a variety of mammals" (2010: 680). Given the evolutionary story of any species, we *expect* other species to share many of its essential properties. According to our essentialist doctrines, what is sufficient for being a zebra is *having all, not just one* of a zebra's essential properties. So Ereshefsky's sufficiency claim is not pertinent. In contrast, his claim that the mechanisms are not *necessary*, that "some zebras lack that mechanism" (2010: 680), is pertinent. But he provides no evidence for the claim! What lies behind it? In particular, what causes the stripes in those zebras alleged to lack the mechanism? Perhaps Ereshefsky has in mind one of the points discussed in section 2.4.

(1980: 250). The essence of *T* is whatever property, as a matter of actual fact, plays that causal role in an environment.[26] This is the basis for the "added metaphysical claim".

This view of essences generalizes to nonbiological kinds. Thus, the essence of gold causes instances of it to be malleable in its normal environment; and see section 2.9 for the causal role of the essences of "artifacts". There is nothing novel or eccentric in thinking that essences must play this causal role. Thus, according to Stephen Boulter, it is "relatively uncontroversial among Aristotle scholars" that Aristotle held that "the essence of a kind has an explanatory role in that it is adverted to when explaining why an instance of the kind has the properties and behavior patterns that it does" (2012: 86).

So Lewens is right in supposing that I do not "wish to say that species have intrinsic essences just so long as there are clusters of properties common and peculiar to them" (2012: 755). But he is wrong to wonder whether my "essentialism amounts to nothing more than the unobjectionable assertion that... intrinsic properties of organisms are relevant to determining their species" (2012: 756). Intrinsic essences play a fundamental causal role.

Essences yield modalities.[27] Obviously, *T* has any of its essential properties *necessarily*. But essences yield more modalities. Take any phenotypic property *P* that a member of *T* has because it is a member of *T*, a property that it shares with all other members of *T* in its environment. Then, *necessarily*, given "the laws of nature" and that environment, *T*'s essence causes its members to have *P*. These necessities, grounded in essences and laws, are necessities in the strong "metaphysical" sense.

In thinking about the criticisms that we have been discussing, one should keep in mind that if *Partly Intrinsic Taxon Essentialism* were wrong and there was no intrinsic component to the essence of a taxon, then the only component would have to be relational. That is indeed the explicit view of some of these critics, as we shall now see. But why should we believe that view? And what precisely is the relational essence? And how does it fulfill its explanatory burden? I shall consider Ereshefsky's answer (also Richards' and Leslie's) in the next section. A strength of the case for *Partly Intrinsic* is the absence of any plausible relational alternative. That case increases if I am right that there is indeed an historical *component* to a taxon's essence but it is one that *demands an intrinsic component*. For, the relevant history of a taxon is of organisms of *a certain intrinsic kind* evolving into

[26] My claim here should not be confused with the following alternative view: the essence of *T* is the property of causing members of *T* to have their shared phenotypic properties in an environment. Thus, suppose that the property that causes the shared properties in the actual world is *P1* but in another possible world *P2* causes a lot of organisms to share those properties. Then, according to my claim, those organisms with *P2* are not members of *T*; according to the alternative, they would be.

[27] But, as Kit Fine has argued persuasively, essences cannot be reduced to modalities: "the assimilation of essence to modality is fundamentally misguided" (1994: 3). Rather, essences are the *source* of modalities.

organisms of *a certain other intrinsic kind*, until we reach the taxon in question (2018a; Chapter 3 below).

2.6 The Irrelevance of the Species Concepts

Ereshefsky poses the taxon problem for species as follows: "why are organisms *O* members of species *S*?" His answer starts on the wrong foot by presenting the following view with approval:

> Relational essentialists argue that modern species concepts posit relational properties, such as interbreeding, genealogy, and occupying a specific niche, as the defining features of species. (2010: 680)

Relational essentialists do indeed argue this but the argument does not deserve approval. For, the species concepts do *not* posit relational properties as defining features of *species S*; they *do not* posit an answer to the *taxon* problem (2008: 356–63, 366–70; 1.5, 1.9 above). Rather, they posit relational properties as defining features of *what it is for S to be a species*; they posit an answer to the *category* problem. Take the popular biological species concept ("BSC"), for example. BSC defines species as "groups of interbreeding natural populations that are reproductively isolated from other such groups" (Mayr 1969: 26). Now, what does that tell us about the property *E* that makes organisms members of taxon *T* and hence constitutes the essence of *T*? Well, we already know that if *T* is a species, then the breeding behavior of its members is one, though only one, of the phenotypic properties that *E* is supposed to explain. So, BSC tells us that *if T* is a species—a big "if"!—*T*'s having *E* must explain why the members of *T*, in their "normal" environment, form a group of interbreeding populations that is reproductively isolated from other such groups. This *is a constraint on* a theory of *E* for species but it *is not* a theory of *E*. Ereshefsky is engaging in the standard practice of conflating the taxon/category distinction 2, described in section 2.3.[28]

Ereshefsky notes my point that "citing the relations among the organisms of a species does not explain why particular organisms are members of a certain species" (2010: 681). So the citing does not explain why this interbreeding group is a group of zebras, that one, horses, and so does not distinguish the essences of

[28] Lewens and Leslie engage in it too. Thus, Lewens claims that BSC and other species concepts "make species membership a matter of relational, rather than intrinsic, properties" (2012: 752). And Leslie, in her positive suggestions about taxon essentialism, starts by following LaPorte (1997; 2004) and Okasha (2002) in favoring relational essentialism because of the species concepts (2013: 138–40). She ends by urging pluralism about relational essentialism because, following Kitcher (1984), she thinks that there are many acceptable species concepts (2013: 140–2).

zebras and horses. We shall return to this in a moment. But, first, it is important not to underestimate what the citing of those relations does not do. Even if *T* is a species, the citing of the relations specified by species concepts like BSC not only does not tell us what *E* is, it does not even tell us *what sort* of property *E* is; it does not tell us, for example, what sort of property does explain the breeding behavior. Indeed, as I point out (2008: 361; 1.7 above), BSC is, on the face of it, quite consistent with *Partly Intrinsic* and the view that intrinsic underlying essence explains that breeding behavior.

Ereshefsky motivates his relational essentialism by appealing to the species concepts. Yet, as I have been emphasizing, those concepts could tell us nothing about the essences of countless taxa that are *not* species. We need to know not only about the essences of the species *Equus zebra* (zebras) and the species *Equus caballus* (horses) but also about the essence of the genus *Equus*. Ereshefsky has nothing to say about the latter: the taxon problems for non-species go entirely undiscussed.

The fact that the species concepts could, *at most*, tell us something about the essences of species is a sign of the irrelevance of these concepts to the essentialism issue. For, the very same issue comes up for any taxon whether or not it qualifies as a species according to some species concept. The examples of the Sumatran tiger and the African forest elephant (2.4.1) help to demonstrate the irrelevance. What is it to be a Sumatran tiger or an African forest elephant? These taxon problems remain whether those taxa are subspecies, as previously thought, or species, as biologists are now inclined to think. Changing our view of the category of these taxa changes a constraint on the answer to the taxon problem for these taxa but otherwise has no bearing on that answer. Here is another example, this time where the reclassification has gone the other way. The British red grouse was once thought to be a distinct species of the genus *Lagopus* but is now classified as a subspecies of the willow grouse and named *Lagopus lagopus scoticus*. This reclassification tells us almost nothing about what it is to be a red grouse.

I have argued that a complete solution to the taxon problem requires more than citing the relations among the organisms of a species, for those relations do not explain what it is to be a member of a *certain* species, hence what it is to be a zebra rather than a horse. Ereshefsky is unmoved: "I believe that the relational essentialists are right that citing relations, without positing intrinsic essences, sufficiently answers the taxon question" (2010: 681). Still, he recognizes that he needs to say more to solve the taxon problem and so continues:

> Here we must turn to relations: particular population and genealogical relations among organisms. The answer to why particular reproductive mechanisms are mechanisms of species S is that those mechanisms occur in organisms whose populations are genealogically connected in a single lineage. (2010: 681)

This is puzzling: Ereshefsky proposes to solve the problem by adding more relations that *obviously* won't distinguish zebras from horses. For, the mechanisms in *both* zebras and horses "occur in organisms whose populations are genealogically connected in a single lineage".

In any case, Ereshefsky soon seems to acknowledge the incompleteness of his solution:

> Devitt could respond that I have not answered the taxon question because though I have bundled intrinsic reproductive mechanisms by citing relations, I have not explained why those relations are relations of a particular species. (2010: 682)

To deal with this objection, Ereshefsky appeals to a particular speciation event:

> At this juncture I suggest that we turn to the particular speciation event of a species. Why are $O_{1...n}$ members of S? Because Os have certain intrinsic reproductive mechanisms that are bound by population and genealogical relations that are anchored to a particular speciation event. (2010: 682)

This proposed solution is still incomplete. First, a particular speciation event often (always, on the Hennig 1966 view) yields *two* new species. We need an account of what distinguishes their essences. Second, we need an account of what distinguishes one speciation event from another. Zebras and horses are (probably) anchored to different speciation events. If those events are to feature in the essences of zebras and horses, the events need to be identified and distinguished. And it will do no good to do this by referring to the different ancestral species involved in the events, for distinguishing the essences of the ancestors poses just the same problem as distinguishing those of zebras and horses.

I mentioned earlier (2.3, 2.5) the failure of the consensus to provide a plausible worked-out relational alternatives to *Partly Intrinsic*. An alternative must be *complete* in that it distinguishes one taxon from another; for example, zebras from horses. And it must be *plausible* in that it posits an essence that can carry the explanatory burden.

Ereshefsky does not clearly address that problem. We noted in section 2.5 that he distinguishes (in effect) structural and historical explanations and then goes on:

> So a robust explanation of why zebras have stripes cites both the relations and intrinsic properties that cause stripes. Merely citing relations provides a relatively weak explanation of an organism's trait. (2010: 680)

Later he says: "while I agree with Devitt that relational essentialism offers weak answers to the trait question, that does not imply that we should adopt intrinsic essentialism" (680). Is he suggesting that relational essentialism *does* offer a structural explanation of why zebras have stripes, albeit a *weak* one? Is he suggesting that a relational essence along with the environment causes each zebra to develop stripes? He does seem to in a later paper where he proposes that "a requirement of being a member of a particular species is having a certain origin or ancestor... and its historical path after that origin" (2014: 11). This is a version of the consensus view that the essence of a species is to be found in its relation to *a certain ancestral taxon* or *certain ancestral individuals*. Leslie (2013: 140, n. 24) and Richards (2010: 169) also make proposals of this sort.[29] So zebras are distinguished from horses in that the particular taxon or organisms from which they originated are different.

I argued in "Resurrecting" that such proposals are explanatorily hopeless (2008: 361–3, 366–70; 1.7 and 1.9 above; also 2018a and Chapter 3 below). In brief, such relational essences could not bear the burden of giving structural, or even historical, explanations of generalizations like that zebras are striped. It is no help to be told that such a generalization holds because the organisms in question are descended from *T**, which has no nature beyond having descended from *T***, which has no nature beyond having descended from *T****, and so on. And, it is no help to be told that the generalization holds because the organisms are descended from Ur-zebras.[30] There is nothing in such purely relational natures that has any causal relevance to having stripes or any other phenotypic property. In brief, such natures cannot meet what I have just called "the Sober demand" (2.5). So far as I can see, neither Leslie, Richards, nor Ereshefsky presents any considerations against this argument.[31]

In conclusion, the species concepts do not provide a motivation for a relational view of a species' essence. Ereshefsky does not address the taxon problem for non-species. His relational essentialism for species in the earlier paper (2010) is incomplete in that it does not distinguish the essence of one species from that of another. And the view in the later paper (2014) still does not address the deep problem for relational essentialism that it cannot fulfill the explanatory burden of an essence.

2.7 The Conspecificity Diagnosis

Why has the significance of distinction 2 between the taxon and category problems been standardly overlooked in discussing essentialism?[32] My "tentative

[29] For other examples, see Ruse (1987: 344); Matthen (1998: 120); Griffiths (1999: 219); Okasha (2002: 200–1); LaPorte (2004: 54).

[30] One might argue, and I have (2018b; Chapter 4 below), that the Ur-zebra has an individual essence that explains why it is striped. Still, this does not save the relational view of a taxon's essence (2018a; Chapter 3 below).

[31] Lewens does not attempt a complete solution to the taxon problem. Slater criticizes *Partly Intrinsic*'s solution, as we noted (2.4). He is also dubious of historical essentialism (2013: 55–61). So, I wonder where he stands on the taxon problem.

[32] I am much indebted to Matt Barker for many detailed comments on the next two sections. These have led to many improvements.

diagnosis" in "Resurrecting" was that this overlooking came from supposing, wrongly, that the species concepts imply relational answers to the *conspecificity* problem, the problem of what makes two organisms conspecific (2008: 363–6; 1.8 above). I have since found a lot of support for this diagnosis, including in Matthew Barker's argument against *Partly Intrinsic* (2010), which will be the concern of this section and the next. So, I am no longer tentative about the diagnosis.

The diagnosis depends on the obvious fact that *Partly Intrinsic*, as an answer to the taxon problem, implies a partly intrinsic answer to the conspecificity problem: if organisms are members of a species partly in virtue of having a certain intrinsic underlying property, then they must be conspecific at least partly in virtue of *sharing* that property. Conversely, if they are conspecific *wholly* in virtue of their relations, then *Partly Intrinsic* is false. And it is tempting to suppose that species concepts do imply wholly relational answers to the conspecificity problem. Thus, BSC implies that individual organisms in an isolated group of interbreeding populations are conspecific with all other organisms in that group. So, it is tempting to suppose that it is *in virtue of* this interbreeding relation that the organisms are conspecific. It then follows that any answer to the taxon problem must be wholly relational. So, people move from a relational view of the **Species** category to a relational view of species taxa via a tempting supposition about conspecificity.[33] That's my diagnosis.

If this diagnosis is indeed right, it is very important to the essentialism issue. For, the diagnosis identifies a crucial but, I argue, mistaken reason for the consensus opposition to *Partly Intrinsic*; the reason that species concepts imply a relational view of conspecificity and hence of a species taxon. So, any defense of this reason is important. Indeed, any defense of the relational view of conspecificity is important. So far as I know, Barker (2010) has given the most thoroughgoing defense. So, the following rejection of this defense, complicated as it surely is, matters to the case for *Partly Intrinsic*.

The argument for *Partly Intrinsic* rejects the tempting supposition of my diagnosis. Thus, even if BSC is true, the interbreeding relations *do not constitute* the conspecificity of those organisms. Rather, the interbreeding relations stand in interesting *causal*, hence *evidentiary*, relations to what does (partly) constitute the conspecificity of members of a group, namely a shared intrinsic state and history (2008: 365–6; 1.8 above). On the one hand, that intrinsic state is central to the *developmental* cause of the interbreeding relations of the group, something adverted to in a *structural* explanation. On the other hand, interbreeding relations are central to the *evolutionary* cause of that intrinsic state in the members of the group; those relations cause gene flow adverted to in the *historical* explanation of the state. Greatly oversimplifying, a shared intrinsic state is the developmental

[33] See, for example, Sterelny and Griffiths (1999), Okasha (2002), and, particularly, Wilson (1999b).

cause of interbreeding relations; interbreeding relations are the historical cause of the shared intrinsic state. It follows, of course, that interbreeding relations provide important evidence of conspecificity.

The point that species concepts do not solve the conspecificity problem is "a bit subtle", as I noted (2008: 365; 1.8 above). It may help to appreciate the point if we generalize it to all categories and taxa.

Inspired by the term 'conspecific', let us call organisms that are members of a taxon *T* that is in a certain category **C**, "con-*C*ic". Now consider these questions:

1. In virtue of what is an organism a member of a taxon *T*?
2. In virtue of what is taxon *T* in category **C**?
3. In virtue of what are organisms con-*C*ic?

Answers to 1 and 2 *jointly* entail answers to 3. For, an answer to 1 tells us what makes an organism a member of taxon *T* and an answer to 2 tells us what puts *T* in category **C**. So, if some organisms are *T*s, according to 1, and *T* is in **C**, according to 2, then those organisms will be con-*C*ic. Taxon essentialist doctrines answer 1. Species concepts answer 2 where **C** is the category **Species**. We need other answers, of course, where **C** is **Subspecies, Genus**, and so on. Suppose that an answer to 2 is that, for *T* to be in **C**, *T*'s members must have property *P*. The key point is that even if we know that *T*'s members have *P* and so *T* is in **C**, this does not tell us that it is in virtue of having *P* that those members are con-*C*ic: *no answer to 2 alone answers 3*. For, answering 3 also requires an answer to 1. Thus, if we know that the members of *T* are in an isolated group of interbreeding populations, then BSC's answer to 2 tells us that *T* is a species. But this alone does not tell us in virtue of what those organisms are conspecific. All we know at this point is that they are conspecific in virtue of *whatever it is in virtue of which they have that interbreeding relation*. Of course an answer to *1 alone* does not answer 3 either. Thus suppose *Partly Intrinsic* is right and taxa we think of as species have partly intrinsic underlying essences. To infer from this what makes organisms conspecific, we would need an answer to 2: we would need to know in virtue of what those taxa are species.

Aside. This has bearing on Leslie's discussion of *chemical* essentialism (2013: 143–52), influenced by Needham, and mentioned in an aside in subsection 2.4.5 above. The passages from Putnam and Kripke that Leslie quotes are primarily concerned with what it is for something to be water or gold (cf. question 1 and the taxon problem). Yet Leslie thinks that it seems "reasonable to infer" from these passages a view of "the *same substance* relation" (p. 143; cf. question 3 and the conspecificity problem), a view that she then goes on to criticize at length. But this inference is only reasonable if Kripke and Putnam have taken a clear stand on what it is for water or gold to be "substances" (cf. question 2 and the category problem). I don't see evidence that they have. Nor, contrary to another critic,

Tuomas Tahko (2015), do I see evidence that they have taken a clear stand on what it is for water or gold to be "chemical" or "natural" kinds. For more on this see Devitt (2021c: 421–3).

Returning to biology, Barker's rejection of the above point about conspecificity is central to his response to my case for *Partly Intrinsic*: he claims that species concepts *do* imply a purely relational answer to the conspecificity problem. Hence they imply a purely relational answer to the taxon problem:

> all prevailing definitions of species concepts provide wholly relational answers to the conspecificity problem, not just to the category problem. Hence those definitions entail wholly relational answers to the taxon problem and are therefore incompatible with Devitt's view.... on any one of these definitions organisms are conspecific in virtue of participating in certain causal processes; definitions are then distinguished by identifying distinct defining causal processes. (2010: 76–7)

These causal processes provide "wholly relational grouping criteria" which "pick out the properties in virtue of which organisms are conspecific"; in the case of BSC, for example "organisms are conspecific in virtue of partaking in reproductive processes, such that they stand in reproductive relations to each other, relations in which nonconspecifics do not stand to each other" (p. 77).

'Criterion', like 'determine' (discussed in subsection 2.4.7), is trickily ambiguous, sometimes epistemic, sometimes constitutive. Barker has *constitutive* criteria in mind here, of course. He is claiming that, according to species concepts like BSC, organisms are conspecific *in virtue of* relational groupings; that's *what constitutes* their conspecificity. That is what I shall reject. Still, BSC does imply that conspecifics *do* form an isolated group of interbreeding populations. Hence, it implies that such interbreeding counts as *evidence* of conspecificity. Hence, on the common *epistemic* meaning of 'criterion', Mayr's definition does indeed imply a criterion for conspecificity.[34] But that is not what concerns us.

Let's call any wholly relational view of conspecificity, like the one Barker describes, "R-CON". *Prima facie*, Barker's confident pronouncement that species concepts provide R-CON is simply false. Take BSC, for example. I earlier (2.6) quoted Mayr's version: "groups of interbreeding natural populations that are reproductively isolated from other such groups". Barker quotes the very same version shortly before making his pronouncement (p. 76). Yet, this species concept provides a "ranking criterion" but *not* a "grouping criterion". It does not tell us what constitutes conspecificity; indeed, it does not even mention

[34] In recent years, there has been increasing interest in the "species delimitation" problem. This is the epistemic problem of *discovering* when a group of organisms form a species, which is sharply distinguished from the "ontological definition of species" (Camargo and Sites 2013: 225).

conspecificity.[35] And it is easy to see that the same goes for all the other species concepts. Indeed, if my generalized argument about con-Cic is right, no species concept alone *could* provide this account: "no answer to 2 alone answers 3". So, in the case of species, *no answer to what makes a taxon a species alone could answer what makes organisms conspecific.*

Yet Barker seems to claim that everyone in the know thinks otherwise: "Among species aficionados it is common knowledge that all prevailing definitions of species concepts propose *answers* to the conspecificity problem" (2010: 76). How could this be? Barker's explanation is that, although species concepts *do not explicitly* provide relational answers to the conspecificity problem, R-CON, these aficionados believe that the concepts *do so implicitly*: "aficionados widely appreciate the implicitness" of grouping criteria in species concepts, the criteria that determine conspecificity (p. 78, n. 5). He supports this claim with a few citations of aficionados who link R-CON somehow to species concepts (p. 77, n. 3; p. 78, n. 5). He might also have cited the works that led me to my tentative diagnosis; see note 33. And we might now add my critics, Ereshefsky (2010: 681), Leslie (2013: 140–1), and perhaps Lewens (2012: 752). So, I'm inclined to think that Barker is right about what aficionados believe. But then the crucial question is: *Are the aficionados right* to believe this? Barker needs to show that their "common knowledge" really is knowledge; that the species concepts really do "propose *answers* to the conspecificity problem", even if only "implicitly".

I anticipated a move like Barker's:

> Now we could, of course, *supplement* BSC as it stands with a relational answer to the conspecificity problem: organisms are conspecific *in virtue of* being able to interbreed. (2008: 366; 1.8 above)

It is trivial, of course, that a species concept supplemented with R-CON provides R-CON. I called such a supplementation "gratuitous" (2008: 366; 1.8 above). And that's the right word for the claim that the supplementation is "implicit" in species concepts. Species concepts, as presented, are straightforwardly "ranking criteria" providing theories of what places a taxon in the rank/category of **Species**. As such, they are largely independent of any theory of conspecificity. Thus, consider BSC:

> BSC's category answer is compatible with [*Partly Intrinsic*]'s answer to the conspecificity problem: it is compatible with the view that organisms are conspecific in virtue of sharing a certain intrinsic underlying property and, perhaps, a history. (2008: 365; 1.8 above)

[35] For more on this, see Devitt (2008: 365; 1.8 above).

There is no basis in linguistic theory for saying that a theory of conspecificity is "implicit" in statements of species concepts.

This is important to the evidential status of R-CON, the relational view of conspecificity. It may seem plausible that the truth about the **Species** category is likely to be found somewhere among the various species concepts. So, had all those concepts implied R-CON, R-CON would be well supported. But there is no basis for saying that the concepts imply R-CON. And the likely fact that aficionados *believe* that the concepts imply R-CON is not adequate support. So, Barker has so far given us no good reason for R-CON.

Still, R-CON might be true, of course. The consensus is that it is true, as Barker emphasizes, and we should expect, given the consensus' relational view of the taxon problem. The consensus might be right to think the R-CON is true even if wrong to think that it is provided by species concepts.

Before considering whether R-CON is true, two comments are called for on my tentative diagnosis. My diagnosis was offered to explain the standard practice in essentialism discussions of overlooking the significance of distinction 2, the distinction between the taxon and category problems. The explanation was that this overlooking came from supposing, wrongly, that the species concepts imply R-CON as the answer to the conspecificity problem. My first comment is that, since I rather agree with Barker that aficionados believe "that all prevailing definitions of species concepts propose *answers* to the conspecificity problem", my attitude to the diagnosis has changed from "tentative" to "firm". My second comment is that, since the aficionados' belief is false, it does not justify the standard practice.

2.8 The Relational View of Conspecificity (R-CON)

It is difficult to keep track of the disagreements between Barker and me over R-CON and *Partly Intrinsic*. The main cause of the difficulty is that the focus of his discussion is on R-CON and the conspecificity problem whereas the focus of mine is on *Partly Intrinsic* and the taxon problem. To cope with this difficulty, it helps to keep firmly in mind the earlier-stated relation between *Partly Intrinsic* and R-CON: if organisms are members of a species partly in virtue of having a certain intrinsic underlying property—*Partly Intrinsic*—then they must be conspecific partly in virtue of sharing that property—not-R-CON. We can capture this relation briefly in either of two ways:

(TC) If *Partly Intrinsic* then not-R-CON; or if R-CON then not-*Partly-Intrinsic.*

"Resurrecting" is a direct argument for *Partly Intrinsic*. So, given the first disjunct of (TC), it is an indirect argument against R-CON. That argument is the case against R-CON that I presented (2008: 366; 1.8 above).

Something else to keep firmly in mind in my exchange with Barker is that *any* proposed solution to the conspecificity problem, including R-CON, *obviously does not alone provide* a solution to the taxon problem. For, any such proposal tells us what it is for two organisms to be members of the *same species* but it does not tell us what it is for them to be members of *one species in particular* (2008: 364; 1.8 above). Thus, the grouping criteria for R-CON that Barker discusses tell us what makes *this* group of animals conspecific, and what makes *that* group of animals conspecific, but they do not tell us what makes the former group zebras and the latter, horses. *Whatever one says about the conspecificity problem, more work has to be done to solve the taxon problem.* What R-CON does provide, given (TC), is a constraint on any such solution: R-CON requires that the solution be wholly relational: for, "if R-CON then not-*Partly-Intrinsic*".

Turn now to Barker's response to my argument for *Partly Intrinsic* and hence against R-CON. Early in his paper, Barker distinguishes my "positive" and "negative" arguments and promises to "argue that neither succeeds" (2010: 74). He rightly identifies as central to my negative argument the argument that "wholly relational answers to the taxon problem . . . are inadequate" (2010: 75). This is my argument (2008: 361–3, 366–70; 1.7 and 1.9 above; also 2018a and Chapter 3 below), just summarized in section 2.6, that such relational answers are "hopeless". They are hopeless because they cannot meet the Sober demand that the essences they posit *carry the explanatory burden*; for example, the stripyness of zebras cannot be explained simply by their being descended from Ur-zebras. Given the importance of this argument to my case for *Partly Intrinsic*, we should indeed expect Barker to attempt to show that the argument does not succeed. Furthermore, we should expect Barker, *as part of that attempt*, to provide a relational answer to the taxon problem that escapes the charge of explanatory hopelessness. Neither of these expectations is met.

The problem with Barker's discussion is present at its very beginning. He conflates my actual argument on the taxon problem with an imagined argument on the conspecificity problem, a conflation of the distinction 3 made in section 2.3 above:

> Devitt suggests that if definitions of prevailing species concepts are interpreted as (in part) answers to the *conspecificity* problem [emphasis added], they are metaphysically hopeless answers to that problem. To argue for this he identifies wholly relational grouping criteria that he thinks may be or have been associated with definitions of prevailing species concepts. . . . Devitt considers the wrong grouping criteria . . . For instance . . . organisms are of the same species iff they are each members of the interbreeding group that contains the type specimen for the species. (2010: 79–80)

The right grouping criteria, in Barker's view, are the ones mentioned earlier, that he alleged are implicit in species concepts, and featured in his version of R-CON.

A clarification. Barker's use of "metaphysically hopeless" picks up on my use of that expression on one occasion (2008: 362; 1.7 above) to describe the *explanatory* hopelessness of relational answers to the taxon question. The occasion was in making a contrast with the possible *epistemic* utility of the answers.

Barker has mistaken the target of my charge of hopelessness. We saw in section 2.6 that the charge is directed at a relational answer to the *taxon* problem. It is not directed at a relational answer to the *conspecificity* problem (R-CON). Its targets are not grouping criteria at all and so cannot be the "wrong grouping criteria", including the "instance" of such a criterion that Barker gives. I argue that wholly relational answers to the *taxon* problem fail the Sober demand and are explanatorily hopeless. I make no similar charge against answers to the conspecificity problem. So, the "wholly relational grouping criteria" that Barker identifies and charges me with neglecting, are not, as he claims, alternatives to what I argued was hopeless; they are proposed answers to a different (though related) problem.

Barker claims: "Prima facie, there seems nothing metaphysically hopeless about the grouping criteria [for conspecificity] I identified" (p. 80). He may well be right. But he is wrong to claim: "The burden is on Devitt to ... show otherwise" (p. 80). My objection to Barker's answer to the conspecificity problem, like to *any* relational answer (R-CON), is *not* that it is explanatorily (metaphysically) hopeless but that it is *false*. Explanations that seem hope*ful* can turn out to be false, as science is constantly demonstrating. The reason I gave for thinking R-CON false was that it is "at odds with" *Partly Intrinsic* (2008: 366; 1.8 above), as it clearly is; see (TC). I had no other reason then, and I have no other reason now. Perhaps Barker's grouping criteria version of R-CON *would otherwise* be a good answer to the conspecificity problem. So, my objection to Barker's version of R-CON is my objection to R-CON in general. And since my objection is *not* that R-CON is explanatorily hopeless, I do not have the alleged burden of showing that Barker's version *is* explanatorily hopeless. My objection to R-CON is that it requires a wholly relational solution to the taxon problem and *that solution* is explanatorily hopeless, as my argument for *Partly Intrinsic* shows (2008: 361–3, 366–70; 1.7 and 1.9 above; also 2018a and Chapter 3 below). So, popular as Barker shows R-CON to be, it entails something quite unacceptable. The burden is then on Barker to show that I'm wrong about this; for example, to show that, contrary to what I argue, (2.6), being descended from Ur-zebras *does* explain the stripyness of zebras.

As a result of this conflation of the conspecificity and taxon problems, Barker does not deliver on his promise to show that my "negative" argument does not succeed. For, he does not respond to my actual hopelessness argument but to his misrepresentation of that argument as aimed at R-CON. Furthermore, as we shall now see, he does not provide what a rebuttal of my actual argument requires: a wholly relational answer to the taxon problem that is *not* explanatorily hopeless.

Barker's failure to give an answer to the taxon problem is important because, as I noted, "as soon as one tries to give one, it should become apparent how explanatorily inadequate a relational answer must be" (2008: 366; 1.7 above).

Early in his paper, Barker nicely distinguishes the conspecificity problem—"In virtue of what are *x* and *y* of the same species?"—from the taxon problem—"In virtue of what is *x* an *F*?" (2010: 75). Yet by the end, the significance of this distinction seems to have been lost. Thus, he restates his answer to the conspecificity problem:

> that (in the right environment) what makes you and me conspecific is that we are causally connected by a network of gene transmissions that have distributed certain genes, dampened the effects of others, and reached reproductive barriers that separate these processes and us from nonhumans.

He then goes straight to the following question about the taxon problem:

> Why think metaphysics demands that *being* human involves something more than appropriately participating in these processes? (p. 81)

And the answer to this rhetorical question is that the reproductive relations that Barker (mistakenly) thinks make him conspecific with a lot of organisms that are humans would also make the late lamented Cecil conspecific with a lot of organisms that are lions. *Barker has not told us what makes him a human and Cecil a lion; his grouping criteria do not distinguish one species from another.* Barker writes as if answering the conspecificity problem answers the taxon problem. But it doesn't, as I emphasized at the beginning of this section. Barker has not presented a wholly relational rival to *Partly Intrinsic.*

Where does Barker's discussion leave *Partly Intrinsic Taxon Essentialism* and R-CON? First, that discussion does not support the apparently common view that the species concepts imply R-CON and hence count against *Partly Intrinsic* via (TC)'s "If R-CON then not-*Partly-Intrinsic*". Second, Barker misrepresents my argument that relational answers to the taxon problem are explanatorily (metaphysically) hopeless and so fails to respond to my actual argument. If that argument stands then, given (TC)'s "If *Partly Intrinsic* then not-R-CON", the argument counts against R-CON. Third, Barker has not proposed an alternative to *Partly Intrinsic.* Fourth, Barker's discussion is entirely about species taxa: the essentialism problems for other taxa go unmentioned. I conclude that *Partly Intrinsic* is the right answer to the taxon problem and R-CON is the wrong answer to the conspecificity problem.

I turn finally to the criticisms of Marion Godman and David Papineau (2020) and then of those two together with Antonella Mallozzi (Godman et al. 2020).

2.9 The Essence of Implements ("Artifacts")

"Resurrecting" (2008; Chapter 1 above) went against the consensus in arguing for *Partly Intrinsic* and hence that the essence of a species is not *fully* historical. Godman and Papineau ("G&P") argue for the consensus position, as the title of their paper indicates: "Species have Historical not Intrinsic Essences" (2020).

My argument for *Partly Intrinsic* has rested on biological explanation. G&P emphasize this too:

> Essential properties are properties that explain all the other shared properties. For any Kind *C*, there will be some central common feature *E* possessed by each *C*, a feature that gives rise to all the other properties *F* shared by the Kind. The essential property thereby explains why the Kind supports multiple generalizations. (2020: 358)

The essence *E* of kind *C* is "super-explanatory" (Godman et al. 2020). So, we are *very much* in agreement on this methodologically significant point. Yet we end up with very different conclusions.

G&P think of species as "historical kinds" which they contrast with "eternal kinds", using terms they take from Ruth Millikan (1999, 2000). They have the good idea of throwing light on species by considering some nonbiological kinds. So, they begin their argument by discussing the essence of a range of such kinds that they think are also "historical". And they claim that I think that species are "eternal". I find these terms quite unhelpful and so will not argue about their application to any kind. My focus in discussing species and some of these other kinds is *very simple*: Do these kinds, whether appropriately called "historical" or "eternal", have an intrinsic component to their essences? For *that* is what is at issue with *Partly Intrinsic*. Whether the essence of a species also has an historical component is not, of course, at issue.

One clear difference between us is that I think, but G&P do not, that essential *intrinsic* properties answer what I call (1.3), following Kitcher, "structural" questions and they call, following Mayr, "proximate" questions. In contrast, G&P think that essential *historical* properties answer those questions (2020: 362). However, in section 2.11, we shall see that G&P's proximate questions differ a bit from mine.

In this section, I shall consider the essence of the nonbiological kinds that G&P discuss, and some others. In section 2.10, I will respond to their criticisms of *Partly Intrinsic*'s view of species essentialism. In section 2.11, I shall take up the "more fundamental objection" that they make in the company of Mallozzi. I shall also look critically at their positive view.

The first nonbiological kind that G&P discuss is a book:

> Consider all the different copies of *Alice in Wonderland*, including the paperback with a front page torn off on Marion's bookshelf, the hardback in David's study, and the many others in numerous libraries and book stores across the world. These instances all share their first word, their second word . . . and so on to the end. They also share the same list of characters, the same plot, and the same locations. We thus have a wealth of generalizations of the form *All copies of Alice in Wonderland are F.* Copies of *Alice in Wonderland* form a Kind. But the common properties of this kind are certainly not explainable by any common physical essence. . . . Rather, all these instances are members of the same Kind because they are all copies of an original. Their shared features are all due to their common descent from the original version written by Lewis Carroll. It is purely this chain of reproduction, not any common intrinsic property, that explains the shared features. (2020: 359)

This is where G&P introduce their talk of *copying*, which is an important sign of their approach to essentialism. And, not surprisingly, copying *does* have a place in discussing the essence of a "copy" of *Alice*. I shall set *Alice* aside for a moment. But here's a quick initial thought: talk of x being essentially a copy of y seems a very unpromising way to reject an intrinsic component to the essence of x. For, *to be* such a copy, x must share the intrinsic properties of y! We shall return to this point (2.11.2).

G&P follow their remarks about *Alice* with some about artifacts:

> Many artefacts are like literary works in this respect. Earlier we alluded to all the features common to Vauxhall Zafiras. . . . But here again the commonalities are not explained by some common intrinsic property. While the Zafiras do have many physical properties in common, none of these is distinguished as the source of all the other common features. Rather their many similarities stem from their all being made according to the same original blueprint. They are constituted as a Kind by their common historical source. (p. 359)

After mentioning some other examples, they sum up their view of historical kinds:

> all examples will involve three central ingredients: 1) the existence of a model, 2) new instances produced in interaction with the model or other past instances, 3) this interaction *causes* the new instances to resemble past instances. A chain of reproduction thus generates the relevant historical relations that ground and explain the Kind. (p. 360)

Now G&P's claim is about "many" artifacts not all artifacts. But it is helpful to start by considering the essence of artifacts in general. And the first thing to note is that among artifactual kinds only the typical "trade-marked" ones like the Vauxhall Zafira or the iPhone are *essentially* artifacts. Consider "generic" artifacts like cars or smartphones.[36] These are, of course, made by us and they are so complicated that it may seem as if they *have* to be made by us. This makes it harder to see what is essential to being one of those things. So, let us consider something much simpler: a paperweight. To be a paperweight an object must *have a certain function*, the function of securing loose papers with its weight. Paperweights often have that function because they are artifacts designed to have it. But they often get that function in a very different way: a perfectly natural object like a stone or a piece of driftwood becomes a paperweight *by being regularly used* to secure papers; its use for that purpose is not just one-off but standard. So, whereas having a certain function is essential to being a paperweight, being an artifact is not. Similarly, being an artifact is not essential to being a doorstop, a hammer, a pencil, a chair, or even a car or a smartphone. Putnam once remarked that chairs might have grown on trees. So might cars and smartphones!

We need a word for these kinds of functional objects. I call them "implements" (2005: 155).[37] So what is essential to an object's being a particular kind of implement is having a certain function. An object has that function in virtue of two properties. First, the object's relation to us or to some other organism: a car was made by us for a certain *purpose* and a nest was made by a bird for a certain *purpose*; a paperweight found on a beach is *standardly used* by us for a certain *purpose*.[38] So relations to organisms are essential to kinds that are implements. But, it is important to note, *not one* of the relations G&P pick out in discussing *Alice* and the Zafira are essential to generic implements: these implements need not be "copies", have a "chain of reproduction", have a "common historical source", be "made according to the same original blueprint", or have "a model".

[36] Stephen Schwartz (1978), responding to Putnam (1975) long ago, pointed out that such objects are not essentially artifacts. I have always emphasized this in my own discussions, on which I draw here (e.g., 2005: 155–6). Oddly, despite Schwartz, much of the discussion of these kinds of objects, alluded to in note 38 below, still goes on as if they were essentially artifacts.

[37] I called them "tools" in an earlier discussion (1997: 246–9). Neither term is perfect for the task. English lacks a word with the meaning we want, one covering not only smartphones, paperweights, hammers, and nests but also the likes of *Alice*, portraits, and statues.

[38] This function essentialism about what I am calling "implements" should not be confused with one about "artifacts", a doctrine that is surprisingly popular, as Tim Juvshik has recently emphasized (2021: n. 2). First, it is obvious that the essence of a kind *being an artifact* is simply being made by humans; that's what it *is* for the kind to be an artifact, nothing more and nothing less! Second, humans make some kinds of things without intending them to have any function at all; Juvshik gives the nice example of doodles. Third, many of the objects of interest to the essentialist in this context are not made by humans; some are found objects, some are made by birds. 'Artifact' is not a term that "carves nature at its joints". Perhaps we should charitably construe much talk of function essentialism about "artifacts" as really being a loose way of referring to the kinds of entities that we are concerned with here, many of which happen to be artifacts.

Importantly, the second property in virtue of which an object has the function of an implement is intrinsic. For, the object must have *any intrinsic property that is needed, or at least thought to be needed,*[39] *to perform that function*. Thus, a paperweight has to have an intrinsic constitution that enables it to secure loose papers. No matter how much we intended something to be a paperweight, it won't be one unless it has that constitution; a feather could not be a paperweight.

An implement of a certain kind has to have the required intrinsic property to begin with but it may be able to *lose* that property and yet still remain that kind of implement: a smartphone that has died is still a smartphone; a car with an exploded engine is still a car. To adapt what Kripke says about individual essences (1980: 114, n. 57; discussed in 4.2 below), we are not concerned with that "temporal question" but with what "(timeless) properties" an implement must have had.[40]

A third property *may* constitute the essence of an implement. Sometimes an implement must have intrinsic properties *beyond those necessary for its function*, properties that distinguish it from other kinds of implement *with the same function*: pencils and pens are both writing instruments but they have different intrinsic essences.

Why should we believe these essentialist claims? They are intuitively plausible, I think, but we must do better than that. As G&P point out, "essential properties are properties that explain all the other shared properties". So we should look to such explanations to support our essentialist claims about implements, as about species. Consider two examples. Why are paperweights useful weapons? Because the essential function of a paperweight requires it to have intrinsic properties (which we could spell out) that make it a good weapon. Why is it easier to erase writing from a pencil than from a pen? Because of the essential intrinsic difference between pencils and pens (a difference we could spell out).[41]

Whereas with biological taxa, we locate essences by looking to explanations provided ultimately by biologists, with implements, we locate essences by looking to explanations provided ultimately by engineers (and the like).

Turn now to G&P's examples. First, the Zafira. These cars are trade-marked implements and so differ from generic cars in that part of their essence is indeed the property of coming from a "common historical source", Vauxhall.

[39] This qualification is necessary because of a neat point made by Beth Preston (2009: 217–18). Beaked plague masks were once produced with the function of protecting doctors from the bubonic plague. It was thought that these masks would perform this function because of the miasma theory of how the disease spread. Since that theory is quite false, the masks did not, indeed could not, perform this function. Nonetheless, the objects in question *were* beaked plague masks. To be such a mask, an object must have whatever intrinsic properties were *thought* to be necessary for its performing its function. (Thanks to Tim Juvshik for drawing my attention to this.)

[40] Thanks to Justin Garson and Tim Juvshik for raising this issue with me.

[41] This discussion of generic implements bears on Millikan's discussion of chairs (1999: 56; 2000: 21).

Furthermore, Zafiras are "made according to the same original blueprint", as G&P say. But, in other respects, G&P are wrong about the essence of Zafiras. First, they are not "copies" (and have no "model"): making a Zafira from a blueprint is not *copying* a prototype; a blueprint of a Zafira is not a Zafira. Of course, a trade-marked car *could* be made by copying a prototype, but that is not how Zafiras, and other cars, are actually made. (Note also that trade-marked implements might not be made at all: found objects might be ingeniously marketed under some name as a kind of, say, "magic rock".) Second, and *very* important, many properties of Zafiras *are* "explained by some common intrinsic property".

A Zafira is essentially a car and so its essence includes all the essential properties of cars. So, it must have the function of a car. So, first, it must be appropriately related to our purposes. Second, it must have, initially at least, all the intrinsic properties essential to functioning as a car; for example, having an engine, brakes, and seats. Something without those intrinsic properties—for example, a paper-weight or a smartphone—*could not* be a car.[42] Third, just as a pencil has an intrinsic essence that distinguishes it from other writing implements, so too does a car have an intrinsic essence that distinguishes it from other vehicles: from a van, truck, bus, pram, golf buggy, etc. Furthermore, a Zafira is a *special kind* of car and so there is even more to its intrinsic essence than to that of a generic car. It has to have *the particular sort* of engine, brakes, and seats peculiar to a Zafira; for example, it has to have the special seven-seat arrangement that is a strong selling point.

How do I know all this? Once again, as with species, we should look to explanation not just intuition, relying ultimately on engineers. (a) Zafiras, like cars in general, are useful for suburban shopping. Why? The explanation is to be found in the intrinsic properties we have just alluded to. And, we should note, the historical fact that Zafiras are made by Vauxhall is irrelevant to that explanation: it would make no difference if the car had been made by Ford or not made at all, just found. (b) According to the advertisement, Zafiras are versatile, easy to drive, comfortable for a family, and disabilities friendly. Suppose they are. Each of these properties of the Zafira would be explained by its intrinsic essential properties. Once again, the relation of Zafiras to Vauxhall is beside the point.

So far, then, the trade-marked Zafira has just the same sort of three-part essence as the generic car: a relation to our purposes and some intrinsic properties that together give the Zafira its essential function as a car: and some other intrinsic properties distinctive of the Zafira. But there is a bit more to the essence of Zafiras. I agreed with G&P that, unlike cars in general, Zafiras must have a "common historical source": they must be made by Vauxhall. Why must they? As we have noted this relational property of Zafiras is irrelevant to the explanations we have

[42] This is not to say, of course, that an object cannot have the intrinsic properties to be more than one kind of implement: an object can be both a wastepaper basket and a doorstop.

been considering so far. But, it is not irrelevant to some other explanations; it may, for example, be part of explanations of the reputation and desirability of Zafiras (think of the fact that an iPhone is made by Apple), of its repair record, and so on. But even with these explanations, essential intrinsic properties are likely to be central.

I also agreed that Vauxhall must have made Zafiras according to an original (at least implicit) blueprint. But this requirement brings together what we have already identified as essential *without adding anything new.* (1) We have just agreed that Zafiras must be made by Vauxhall. (2) To say that they must be made according to that original blueprint is just to say that they must be made with the properties specified by the blueprint. *And those properties include the earlier-mentioned essential intrinsic properties of the Zafira*, the ones that carry the burden of explaining such commonalities of Zafiras as its special seven-seat arrangement. In brief, to say that the essence of Zafiras is to be made by Vauxhall according to the original blueprint is not to deny that the essence has an intrinsic component; rather it is to *entail* that it has. The essence that G&P propose is up to its ears in intrinsic properties.

I emphasize that we need to explain the *particular commonalities* of Zafiras—the seven-seat arrangement and so on—not simply that they have *some commonalities or other*. I have argued earlier (2.6) that it is explanatorily hopeless to think that the particular commonalities of a species can be explained by a purely relational essence. The same goes for the particular commonalities of Zafiras.

The story for *Alice* is, as G&P say, much like that for the Zafira. But G&P have both stories wrong. It is essential to a copy of *Alice* that it is a copy of the original manuscript produced by Lewis Carroll just as it is essential to a Zafira that it was made from Vauxhall's blueprint. But intrinsic properties are essential in both cases. For something to be a copy of *Alice*, it has to be a *linguistic* item with near enough the same *semantic* properties as Lewis Carroll's manuscript. That is what it *is* to be a copy and anything without such properties is simply not a copy of *Alice*. And having those properties is partly an intrinsic matter. It is those partly intrinsic linguistic properties – the "first word... second word... list of characters... plot, and locations" – that explain such commonalities as that copies of *Alice* show great insight into the difference between quantifiers and names. No historical essence can explain those commonalities.[43]

Finally, G&P mention another artifact, the pound coin: "We don't think all the pound coins pressed from some mould must have some common inner essence to explain why they share their many other joint properties" (2020: 365). On the contrary, a common inner essence is necessary to explain their behavior in ticket machines and many other properties. Of course, the *most important* properties of

[43] My discussion of blueprints and copying bears on Millikan's discussion of reproduction and copying in discussing the nature of kinds (1999: 54–6).

the coin are its function of being worth one pound, which it has because of its relation to the Bank of England. Still, there is an intrinsic component to its essence.

2.10 Godman and Papineau against *Partly Intrinsic Taxon Essentialism*

After a brief discussion of some other kinds, G&P finally turn to biological essentialism (2020: 360). Their argument against *Partly Intrinsic* to this point has been the suggestion that species are like other kinds, ones they classify as "historical", in not having an intrinsic component to their essence. I think that they are right to look to other kinds for guidance. But what we should learn from studying these others, supported by my discussion of Zafiras, *Alice*, and the pound coin, is that *it is unlikely that the essence of any explanatorily interesting kind is wholly relational.* (*Being Australian* is my favorite example of an explanatorily *un*interesting kind that is probably wholly relational; 2008: 346; 1.1 above.) In any case, in the end, the rejection of *Partly Intrinsic* demands arguments about species themselves not about other kinds. We need to be shown that the structural explanations of species commonalities do not rest on intrinsic properties. So that is what we now look for.

G&P contrast their position on species with mine:

> Why do their members all share so many properties? As we have seen, one answer would be to assimilate species to eternal Kinds, as Devitt does, and appeal to the common genetic make-up intrinsic to each member. But an alternative would be to view species as historical Kinds, and attribute their shared properties to their common ancestry, with their genetic make-up simply being part of the species' copying mechanism. (2020: 363)

My examples of the sorts of shared phenotypic properties that we are talking about include: "ivy plants grow toward the sunlight ... polar bears have white fur ... Indian rhinoceri have one horn and Africa rhinoceri have two" (2008: 351; 1.3 above). And my claim is that an essential intrinsic underlying property of the kind in question is central to the structural explanation of these commonalities, to the explanation of why each of these organisms develop to have the property specified for its kind. That intrinsic component of the essence, together with the organism's environment, causes the organism to have the specified property (2008: 352; 1.3 above).

G&P give two reasons for rejecting my view and for taking species to be "fundamentally historical". The second of these concerns microbial kinds. I did not address such kinds at all in "Resurrecting" and I have acknowledged that

Partly Intrinsic may not apply to them (2.2, n. 6). So I shall set the second reason aside and attend only to the first reason, "non-zygotic inheritance". Does it show that *Partly Intrinsic* is wrong about the non-microbial biological world? I think not. Indeed, I'm puzzled that they think that it does.

G&P start their discussion of non-zygotic inheritance by pointing out that "the children of a skilled forager" may not inherit their skills through their genes but rather

> in other ways – for instance, by her explicitly training them, or by their implicitly copying her tricks. In this sense, there is nothing at all problematic about the inheritance of acquired characteristics. (2020: 363)

Indeed, there is not, and such inheritance is common in nature, as G&P bring out nicely. *But this throws no doubt on Partly Intrinsic.* For, the training and copying are part of the environment's causal role in the development of phenotypic properties. As G&P note (2020: 364), I emphasize the obvious fact that "explanations will make some appeal to the environment" as well as to intrinsic essences (2008: 352; 1.3 above).

At this point, I'm sorry to say, G&P seem to go right off the rails: "nothing requires characteristic traits shared by species members to depend on genetic inheritance *at all*" (2020: 364). Surely they can't really mean "*at all*"?! But they do:

> Why do all tigers grow up the same, and different from zebras, even though tigers and zebras are subject to just the same environmental influences? What could explain that, except their shared genetic make-up? Well, the answer is that tigers and zebras aren't subject to just the same environmental influences. Tigers are raised by tigers, while zebras are raised by zebras, and many of their species-characteristic properties can be due to this in itself – without any assistance from their genes. (p. 364)

Without *any* assistance?! Let's not beat about the bush: this is A Big Mistake. If it weren't for their shared genetic make-up, no tigers would acquire *any* traits from interaction with their parents (or with anything else). If G&P were right, a zebra brought up by tigers would have all the traits that tiger cubs acquire from interaction with their parents. That is surely not so. Chimps brought up by humans famously fail to learn a human language. (Indeed, language acquisition is a good example of the combined action of genes and environment in acquiring a trait.) Young cuckoos don't grow up like their foster parents.

So I see nothing in non-zygotic inheritance that counts against *Partly Intrinsic*. Aside from that, what about all the other commonalities of species? What explains why Indian rhinos have one horn and Africa rhinos have two? Or why tigers and zebras have stripes? These are certainly not traits acquired from watching parents.

No case has been presented against the view that an intrinsic essence provides the structural explanation of these traits.

2.11 The Historical Species Essentialism of Godman, Mallozzi, and Papineau

Godman and Papineau soon joined with Antonella Mallozzi ("GMP") to raise "a more fundamental objection" to my position on essentialism:

> It is not just that this position fails to deal with non-genetic inheritance and non-sexual reproduction. Rather, even if we stick solely to genetically inherited traits in sexual species, it never offers the right kind of super-explanation at all.
> (Godman et al. 2020: 324)

In elaborating this objection GMP develop G&P's sketch of a positive view of species essentialism. This positive view is novel and interesting. It is also, I shall argue, quite inadequate. I shall start with the objection and then move to the positive view.

2.11.1 The "More Fundamental Objection"

It was clear from the start that G&P's proximate explanations had somewhat different explananda from mine. I took mine from Mayr but followed Kitcher in calling them "structural" rather than "proximate" (1.3). I am looking for an essence that provides developmental explanations of generalizations about the phenotypic properties of a species. G&P put this aptly:

(1) Why do the members of a species each develop their range of shared phenotypic properties? (p. 361)

G&P's proximate question is interestingly different. They are looking for an essence that provides an explanation of "a great number of commonalities" and "multiple similarities" in the phenotypic properties of a species (p. 362). What they want explained about species is: "Why do their members all share so many properties?" (p. 363). We can put G&P's question thus:

(1*) Why do the members of a species each *have in common* their range of shared phenotypic properties?

They take this to demand "a common cause" of the shared properties (Godman et al. 2020).

Mayr raised another question which he called "ultimate" (I followed Kitcher again in calling it "historical"). G&P express it thus:

(2) What led to there being a species whose members develop this range of phenotypic properties? (2020: 361)

GMP are explicit that their question is neither (1) nor (2): "Our analysis of biological taxa is motivated by a question that is quite distinct from both of Mayr's questions" (Godman et al. 2020: 325). So where do they stand on (1) and (2)? We shall return to that.

GMP's "more fundamental objection" to my *Partly Intrinsic* is that it does not provide the common cause demanded by (1*):

> the genomic material common to the members of a taxon will typically not be suited to play this role. This is because it will normally be a conjunction of different genetic properties, each one of which explains a different phenotypic feature. So we do not have one intrinsic property acting as a common cause for many phenotypic properties, but simply a list of different intrinsic genetic properties explaining different phenotypic properties. And this thus leaves us once more with an unexplained multiple correlation—why are all those different intrinsic genetic properties found together? (Godman et al. 2020: 324)

I agree. We need to explain not only each generalization we make about the phenotypic properties of a taxon but also the correlation of these phenotypic properties in the taxon. And whereas, I have argued, *Partly Intrinsic* explains the former, it does not alone explain the latter. This is not surprising, of course, because it was not proposed to answer (1*) but just (1). To give the explanation demanded by (1*), we must explain the particular correlation of *intrinsic genetic* properties that explains the particular correlation of *phenotypic* properties; we do indeed need to explain "why are all those different intrinsic genetic properties found together".

For that we turn to the historical component of a taxon's essence. G&P rightly claim: "Members of a species owe their shared properties to their common ancestry" (2020: 361). But it is important to see that this is just a *gesture* toward an adequate explanation: it needs to be filled out. To make it adequate, I look to the already foreshadowed (2.5) conclusion in Chapter 3 (and in 2018a): the essence of a taxon has a certain historical component. This component is of organisms of *a certain intrinsic kind* evolving into organisms of *a certain other intrinsic kind*, until we reach the taxon in question. This provides the explanation (1*) demands. The shared essential underlying intrinsic properties of the common ancestors explain, by an evolutionary process, the particular correlation of shared essential underlying intrinsic properties of the given taxon; and the latter correlation explains the particular correlation of shared phenotypic properties of the taxon. Thus, we meet GMP's demand to explain "why are all those different intrinsic genetic properties found together". And, importantly, rather than undermine *Partly Intrinsic*, that essentialist doctrine is part of the answer to (1*).

I turn now to the positive view of GMP, their historical species essentialism.

2.11.2 The Positive View

How do GMP meet the demands of (1*)? How do *they* fill out their gesture to yield an adequate explanation? We saw the importance of these questions in our earlier discussion of Zafiras (2.9). We noted that a gesture toward a "common historical source" is not an adequate account of the essence: "we need to explain the *particular commonalities* of Zafiras—the seven-seat arrangement and so on—not simply that they have *some commonalities or other*". In particular, the explanation must be "complete" in that it distinguishes Zafiras from Land Rovers (cf. 2.6). And filling out the gesture for Zafiras showed that their essence includes many intrinsic properties.

An analogy offered by G&P helps to make the point about what is needed:

> consider a request for an explanation of the properties common to all Christians. We would say that the members of this Kind display so many common features because they are all influenced by a common historical source. This seems the right answer. But it doesn't require us to be specific about when Christianity started. Maybe we should date it from Jesus, or from the first Pope, or from the Council of Nicea. But our historical account of the Kind would seem to stand up perfectly well whichever we do. (2020: 362–3)

It is true that G&P's account need not give the specifics of the common source; it can leave that to future empirical investigation. Similarly, my account, *Partly Intrinsic*, had no need to give the specifics of the essence of, say, tigers: it left that to science (2.4.5). But any good account must tell us *what sort* of property is the essence. And that sort must be a sort that explains the *particular* commonalities that distinguish Christians from Jews, Muslim, etc. That is the Sober demand on an essence (2.5). And, given what has gone before, we can anticipate that any wholly relational sort of essence is unlikely to meet this demand: it will be explanatorily hopeless.

So far as I can see, G&P (2020) do not move beyond the gesture to provide the explanation that (1*) demands. But GMP (2020) say this:

> To find a property that explains why all the different phenotypic features of a biological taxon are instantiated together, we need to turn away from ontogenetic development and view the taxon as a Historical Kind. The reason the phenotypic

> features of a taxon are found together is that its members are all descended from common ancestors who had those features. Biological reproduction is a copying process—offspring share the heritable features of their parents. So a set of properties that is conjoined in an ancestral population will be found together in the descendants, too. (2020: 324–5)

Now the talk of "copying" here is quite a simplification, as GMP well appreciate (p. 325, n. 11): since a species has *evolved* from its ancestors its members have a *different* range of shared phenotypic properties from those ancestors. But let's go with the simplification. What, on this story, explains the co-instantiation of the particular range of features shared by members of the species? Answer: *the co-instantiation of those features in ancestors*. It is because the species members are "copied" from ancestors with those particular features, and not with any other features, that the members co-instantiate that range of features. So, on this story, the explanatory essence of the species is having ancestors with those features. And having those features is partly an intrinsic matter. Indeed, we foreshadowed such a finding in responding to G&P's first mention of copying in discussing *Alice* (2.9). So, if species members are essentially copies of their ancestors, then *they must share certain intrinsic properties with those ancestors*. That's the way it is with copying! In brief, once GMP's story is filled out, it is not a story of a wholly historical essence.

Consider this earlier passage from GMP:

> it is no explanation of the many features common to all horses, say, to specify that horses all have some genetic material that determines manes, and some other genetic material that determines uncloven hooves, and some other genetic material that determines their head shape, and so on. We would not yet have explained why all these different genetic features are found together in horses. (p. 324)

I have agreed. But, *we don't provide the missing explanation by simply adding that those horses have common ancestors*. This is so incomplete that it wouldn't even distinguish horses from zebras or, indeed, from any taxon. The best way to make this explanation adequate (on the simplification) would be to add that *those common ancestors* all "have some genetic material that determines manes, and some other genetic material that determines uncloven hooves...", and all this genetic material has been passed on to horses via reproduction ("copied"). Without some such addition, we have no explanation of horses having that particular range of common features, a range that distinguishes them from zebras. And with that addition, GMP's essentialism would have collapsed into something like mine: a combination of *Partly Intrinsic* with the partly historical doctrine to be argued for in Chapter 3.

In sum, GMP's answer to (1*), as it stands, is an inadequate explanation of why a particular range of phenotypic properties are shared by members of a species. To make it a good explanation, it should be filled out in a way that entails *Partly Intrinsic*. Indeed, this should hardly be a surprise: one cannot explain why members *share* certain properties without explaining, as *Partly Intrinsic* does, why members *have* each of those properties.

As noted, GMP are explicit that they are not trying to answer Mayr's proximate question (1) or ultimate question (2). Still, they have quite a bit to say that bears on those "traditional" questions (pp. 325–6). They sum up:

> our approach simply assumes answers to the traditional questions, taking as given both the machinery that allows offspring to resemble their parents, and the original traits of founding populations. (p. 326)

I see the following underlying truth here: once GMP's answer to their question (1*) is filled out appropriately, it amounts to an essentialism that entails partly intrinsic and partly historical essentialist answers to the traditional questions (1) and (2).

GMP's story faces one final large problem. GMP take the historical essence of a species to be that "its members are all descended from common ancestors" (2020: 325). What do they have in mind as the "common ancestors"? A later discussion of tigers clarifies: the ancestors that GMP seem to have in mind are "the original tigers" (p. 327). This would make the *actual animals* in the "founder population" essential. That is very implausible, as I have argued (2018a: 3–4), and as we will see in section 3.3 below.

To conclude my discussion of G&P and GMP, there is certainly a relational, sometimes historical, component to the essence of what I have called "implements". But there is also an intrinsic component which plays a central role in the explanation of commonalities. And the same goes for biological taxa. *Partly Intrinsic Taxon Essentialism* stands untouched.

2.12 Conclusion

In "Resurrecting Biological Essentialism" (2008) and Chapter 1, I argued for *Partly Intrinsic Taxon Essentialism*: the essence of a biological taxon consists partly in intrinsic underlying, and probably largely genetic, properties. *Partly Intrinsic* flies in the face of the consensus in philosophy of biology, particularly for species, that the essence is wholly historical and relational. In the present chapter I have defended my paper from the interesting criticisms of Wilson et al. (2007), Barker (2010), Ereshefsky (2010), Richards (2010), Lewens (2012), Leslie (2013), Slater (2013), Godman and Papineau (2020), and Godman et al. (2020).

I had to make some hard choices in responding to my critics and so chose to make detailed responses only where those responses seem to me to advance the case for *Partly Intrinsic* rather than labor the case already made. Distinctions 1 to 3 below (2.3) were central to my responses.

1. The Distinction between Structural and Historical Explanations. Structural explanations of biological generalizations about phenotyptic features of a taxon tell us about the intrinsic underlying states that *make the generalizations true.* Historical explanations tell us how the members of the taxon *came to have* such states. My argument in "Resurrecting" was that structural explanations demand (partly) intrinsic essences.

I have answered criticisms that *Partly Intrinsic* is at odds with certain biological variations (2.4). These criticisms come from Leslie, Wilson et al., Slater, Lewens, and Richards. In my answers, I have emphasized several points. (1) *Partly Intrinsic* gets support from the role of genetic analysis in taxonomy. There is a pattern of tying a taxon to an underlying genetic structure, a structure that causes its phenotypic properties, and of reclassifying where there is significant genetic difference among two populations within a taxon. (2) *Partly Intrinsic* leaves it to biology to tell us precisely what the underlying essence of a taxon is and has no commitment to it being "simple". (3) The essence plays its causal role across *all* environments, the "normal" and the "abnormal".

I responded to the charge by Ereshefsky, and before him, Godfrey-Smith, that *Partly Intrinsic*'s talk of intrinsic essences is an uncalled for metaphysical addition to biology (2.5). The essence or nature of a taxon simply *is* whatever underlying state causes members of that taxon, in their environment, to have their phenotypic properties; the essence explains the place that those organisms have in the causal nexus *just because they are members of the taxon.*

2. The Distinction between the Taxon and the Category Problems. The main objection of the consensus to *Partly Intrinsic* is that the species concepts show that the essence of a species is relational not intrinsic. I have argued that this objection is seriously mistaken, conflating the "category" and the "taxon" problem for species. The *category* problem in general is concerned with what it is for a taxon to fall under a certain category. The species concepts propose answers to this question for the category **Species**. The *taxon* problem in general is concerned with what it is for an organism to be a member of a particular taxon. This is the essentialism problem to which *Partly Intrinsic* proposes an answer. The only bearing that the species concepts have on this problem is to provide a *constraint* on an answer to the taxon problem *for any taxon that is a species.* But they do not provide an answer to the problem for species and are irrelevant to the problem for other taxa.

I argued that Ereshefsky's appeal to the species concepts in objecting to *Partly Intrinsic* exemplifies the standard practice of conflating the distinction between the category and taxon problems (2.6). The species concepts do not support a relational view of a species' essence. Aside from that, the relational view that Ereshefsky proposes in one paper (2010) is incomplete in that it does not distinguish the essence of one species from that of another; and that in another (2014) does not address the deep problem that a relational essence cannot carry the explanatory burden and hence meet the Sober demand. Finally, Ereshefsky's discussion of taxon essentialism exemplifies another standard practice: it is all about species and does not address the essentialism issue for other taxa at all.

3. *The Distinction between the Conspecificity Problem and the Taxon Problems.* *Why* has the significance of distinction 2 been standardly overlooked? My "tentative diagnosis" in "Resurrecting" was that this significance has been overlooked because of the appealing, but mistaken, idea that the species concepts imply "R-CON", a relational answer to the *conspecificity* problem. Take BSC, for example. BSC implies that an individual organism in an isolated group of interbreeding populations is conspecific with all other organisms in that group. So, it is tempting to suppose that it is *in virtue of* this interbreeding relation that the organisms are conspecific. It then follows that any answer to the taxon problem must be wholly relational, R-CON. So, people move from a view of the **Species** category to a view of species taxa via a tempting thought about conspecificity.

Barker's argument embraces this move with enthusiasm. In support, he claims that *aficionados believe* that the species concepts entail R-CON. I'm inclined to think he is right about that and so changed my diagnosis from "tentative" to "firm". But Barker does not show that the aficionados are *right* to believe this and I have argued that they are not (2.7). Even without the support of the species concepts, R-CON might be true anyway, of course. I have argued for *Partly Intrinsic* and against relational answers to the *taxon* problem, claiming that they are explanatorily hopeless (2.6). Now if I'm right about *Partly Intrinsic*, then the relational answer to the *conspecificity* problem, R-CON, must be false too. In his response, Barker misrepresents my argument of explanatory hopelessness and so does not address the actual argument. The argument stands (2.8).

I turned last to Godman and Papineau. They argue that species are historical not intrinsic kinds. They have the good idea of throwing lights on species by considering some other kinds, particularly "artifacts" or, as I prefer to call them, "implements". They argue that these kinds are historical in that they are all, in one way of another, "copies of an original"; they have a "common source". In contrast, I argued (2.9) that copying has nothing to do with the essence of a generic instrument like a car (and, one might add, very little to do with that of a trademarked instrument like a Zafira). Rather, the essence is partly relational and partly intrinsic: it includes a relation to our purposes and the intrinsic properties necessary to function as a car; and it includes the intrinsic properties that distinguish a car from other vehicles. How do we know all this? As usual, we look to what explains the common properties of the kind.

So. I resisted Godman and Papineau's suggestion that a consideration of these other kinds supports an historical view of species. In discussing species directly, they give two reasons for rejecting *Partly Intrinsic* and for taking species to be "fundamentally historical". One concerns microbial kinds, which I am setting aside. The other concerns "non-zygotic inheritance". I found nothing in this inheritance that counts against *Partly Intrinsic* (2.10).

Finally, Godman and Papineau join with Mallozzi to raise "a more fundamental objection". They rightly charge that *Partly Intrinsic* alone does not explain why a certain range of phenotypic properties are *co-instantiated* in the members of a species. For that, I argued, we do need to *add* a certain historical component to a taxon's essence. But *Partly Intrinsic* remains part of the explanation. Furthermore, the alternative explanation that Godman, Mallozzi, and Papineau offer is inadequate as it stands and, once made adequate, ceases to be wholly historical.

In rejecting *Partly Intrinsic Taxon Essentialism* the consensus is committed to a relational answer to the taxon problem. *But what answer in particular?* I have emphasized three requirements on an answer. First, it must be *complete* in that it distinguishes the essence of one taxon from that of another. Second, it must posit an essence that can carry the explanatory burden: the posited essence, along with the environment, must provide *plausible* structural explanations of biological generalizations. Third, it must be *comprehensive* in that it covers taxa other than species. My relationist critics do not meet these requirements. Lewens, Barker, and Slater do not offer a complete answer, even for species. Leslie, Richards, Ereshefsky, and Godman, Mallozzi, and Papineau offer ones but they are explanatorily implausible; indeed, they are "hopeless". Many do not offer a comprehensive explanation: the problems for non-species are not addressed by Ereshefsky, Richards, Leslie, and Slater, and not even mentioned by Lewens and Barker.

In the next chapter, I shall argue that the consensus as a whole has strikingly failed to produce an answer to the taxon problem that is complete, plausible, and comprehensive.

3
Historical Biological Essentialism

3.1 Introduction

What is it *to be* a member of a particular taxon?[1] *In virtue of what* is an organism say a *Canis lupus*? What *makes* it one? As Kim Sterelny and Paul Griffiths make clear in their textbook (1999: 8, 186), the consensus answer in the philosophy of biology, particularly for taxa that are species, is that the essence is not in any way intrinsic to the members but rather is wholly relational, particularly, historical. Samir Okasha endorses the consensus, describing it as follows: we "identify species in terms of evolutionary history... as particular chunks of the genealogical nexus" (2002: 200).[2] Philosophers of biology like to emphasize just how different their historical essentialism is from the influential views of Saul Kripke (1980) and Hilary Putnam (1975).

So, the consensus is:

> *Fully Historical Taxon Essentialism*: Biological taxa have essences that are not intrinsic but wholly relational, particularly, historical properties.

This raises two questions. (A) Why believe it? (B) What *precisely* is this wholly relational essence? I shall address these questions in turn.

In "Resurrecting Biological Essentialism" (2008), and in Chapters 1 and 2, I rejected the consensus. I presented an argument that there is an intrinsic *component* to a taxon's essence and responded to a range of objections to such a view. In Chapter 2, I called this doctrine, "*Partly Intrinsic Taxon Essentialism*". So, my opening response to (A) is that we should *not* believe *Fully Historical* (for short). Still I accepted in "Resurrecting", without argument, that there was also an historical component to a taxon's essence:

> *Partly Historical Taxon Essentialism*: Biological taxa have essences that are partly historical properties.[3]

[1] This chapter modifies Devitt (2018a). Some of its lines of thought overlap with some in Elder (2008).

[2] For others, see Hull (1978: 313); Sober (1993: 148–50); Matthen (1998: 120); Griffiths (1999: 219–22); Millikan (2000: 19); Ereshefsky (2001: 209); LaPorte (2004).

[3] In Devitt (2018a), the doctrines now named "*Fully Historical Taxon Essentialism*", "*Partly Intrinsic Taxon Essentialism*", and "*Partly Historical Taxon Essentialism*" were named, respectively, "*Historical Essentialism*", "*Partly Intrinsic Essentialism*", and "*Partly Historical Essentialism*". I've changed these names to distinguish each doctrine more clearly from other doctrines in this book.

Biological Essentialism. Michael Devitt, Oxford University Press. © Michael Devitt 2023.
DOI: 10.1093/oso/9780198840282.003.0003

This doctrine needs an argument just as does the consensus *Fully Historical*. I offer an argument in section 3.2, analogous to my earlier one for *Partly Intrinsic*, and drawing on the literature: the historical component is required by historical/ evolutionary explanations. Clearly, if this argument is good, and if the consensus were right in rejecting *Partly Intrinsic*, then the argument would establish *Fully Historical*. That ends my discussion of (A).

But most of the chapter is concerned with (B). In Chapter 2, I have emphasized that a relational answer to the taxon problem must, first, be *complete* in that it distinguishes one taxon from another; for example, zebras (*Equus quagga*) from horses (*Equus ferus*); and a species with descendants from its descendants. (I take it that the latter issue does not arise for the higher taxa which can only end when all their descendants are extinct.) Second, an answer must be *plausible* in that it posits an essence that can bear the explanatory burden. In Chapter 2, our focus was on the burden of *structural* explanations. But our concern now broadens to include *historical/evolutionary* explanations. Despite asking around, I have been surprisingly unable to find a worked out complete and plausible answer in the literature. In sections 3.3–3.4, I argue against the two answers I have been able to find. The signs are that we should move away from the consensus *Fully Historical Taxon Essentialism*.

There is a third requirement on answers to the taxon problem. Our concern is with the essence of taxa thought to be in any one of the Linnaean categories but discussions of this are dominated by talk of taxa that are thought to be species. I shall go along with that practice. Nonetheless, an answer to the taxon problem must be *comprehensive* in that it covers taxa other than species.

In "Resurrecting", in accepting that there was an historical component to a taxon's essence, I aired the suspicion

> that even historical explanations demand a partly intrinsic essence; that, for example, the explanation of how polar bears came to be white will ultimately depend on essential intrinsic properties of polar bears and of their grizzly ancestors. (2008: 354–5; 1.3 above)

In section 3.5, I develop this suspicion by arguing that the history relevant to the essence of a taxon is of organisms of *a certain intrinsic kind* evolving into organisms of *a certain other intrinsic kind*, until we reach the taxon in question. The consensus is right that there is an historical *component* to the essence of a taxon—*Partly Historical*—but that component requires that there also be an intrinsic component—*Partly Intrinsic*. So the essence is not *wholly* relational: *Fully Historical* is mistaken.

Why has the dependence of the historical component of the essence on an intrinsic component been missed? I suspect that it is because of the lack of attention to the details of *Fully Historical Taxon Essentialism*.

A final preliminary point. The "working assumption", emphasized in section 2.2, is that the Linnaean taxa that concern our essentialism are "very largely explanatory".

I turn now to my response to (A).

3.2 An Argument for *Partly Historical Taxon Essentialism*

Why believe *Partly Historical*? I shall offer an argument, analogous to my earlier two-part argument in "Resurrecting" for *Partly Intrinsic* (2008: 351–5; 1.3 above). I shall find signs of this analogous argument in the literature.

The first part of that earlier argument concerned the biological generalizations about the phenotypic properties of species and other taxa; generalizations about what they look like, about what they eat, about where they live, about what they prey on and are prey to, about their signals, about their mating habits, and so on. I argued that these generalizations have explanations that advert to an intrinsic underlying, probably largely genetic, property that is part of the essence of the taxon. If we put together each such underlying property that similarly explains a generalization about a taxon, then we have the intrinsic part of its essence. The explanations that featured in the argument were what Mayr called "proximate" and I, following Kitcher (1984), called "structural". They concern the underlying developmental mechanisms in members of a taxon that make the generalizations true. Consider this example: Indian rhinos (*Rhinoceros unicornis*) have just one horn. I argued that there has to be something about the very nature of these rhinos that causes them, in their environment, to develop from embryos into animals with one horn. And that something is part of the essence of the rhinos.

But now consider the other side of Mayr's distinction, "ultimate" explanations that I, again following Kitcher, called "historical". They tell us how there came to be that taxon in the first place, how it evolved. What *led to* there being taxa with the phenotypic properties that are the subject of the generalizations? In light of the argument from structural explanations to an intrinsic component in a taxon's essence, we might expect to find an analogous argument from historical explanations to an historical component in the essence. How did it come about that there are Indian rhinos? What explains the existence of these organisms with developmental mechanisms that cause them to have one horn? There has to be something about the rhinos' evolutionary history that explains this. The ancestors of the rhinos, in their environment, caused the rhinos to evolve with one horn. Having that history is part of the very nature of Indian rhinos, part of their essence. Having a different history caused Black African rhinos (*Diceros bicornis*) to grow two horns and is part their essence. What is it in these histories, precisely, that constitutes these historical essences? This is question (B), to be considered in the sections to follow.

Why do Indian rhinos have one horn? The question is ambiguous. Understood structurally, it demands an explanation of what it is about each rhino that causes it, in its environment, to grow one horn. Understood historically, it demands an explanation of what caused there to be any animals of this sort, with their one horn, in the first place. The structural explanation yields the intrinsic component of the essence, the historical, the historical.

I take it that Griffiths (1999) has an historical explanation in mind when he argues as follows for *Fully Historical Taxon Essentialism.* Quoting Darwin's claim (1859) that "unity of type is explained by unity of descent", Griffiths points out that

> there is a well-known Darwinian ground for expecting groups defined by common descent to share morphological and physiological characters... The principle of heredity acts as a sort of inertial force, maintaining organisms in their existing form until some adaptive force acts to change that form. This *phylogenetic inertia* is what licenses inductions and explanation of a wide range of properties... using kinds defined purely by common ancestry. (1999: 219–20)

In the second, related, part of my argument for *Partly Intrinsic,* I claimed that a taxon's intrinsic essence explains *why* being in the taxon is explanatory:

> the generalizations we have been discussing reflect the fact that it is *informative* to know that an organism is a member of a certain species or other taxon: these classifications are "information stores" (Sterelny and Griffiths 1999: 195). But being a member of a certain taxon is more than informative, it is *explanatory.* Matthen points out that "many biologists seem committed to the idea that something is striped *because* it is a tiger" (1998: 115). And so they should be: the fact that an individual organism is a tiger, an Indian rhino, an ivy plant, or whatever, explains a whole lot about its morphology, physiology, and behavior. (2008: 352; 1.3 above)

Why does it? Because the essential nature of a taxon, to be discovered by biologists, causes its members, in their environment, to have those phenotypic properties. What nature? I argued that if our concerns are structural, so they are with a nature that causes a tiger's development into an organism with those properties, the nature must be intrinsic. But now our concerns are historical, with the nature that led to there being organisms with those developmental mechanisms. The nature that explains that must be historical. For an organism to be a tiger (*Panthera tigris*) is for it to have the historical essence of tigers. Some aspect of the history of tigers, constituting that historical component of their essence, in combination with the environment, brought it about that there are tigers with their stripes and other phenotypic features. That's why being a tiger is explanatory.

In sum, the same historical nature or essence that (partly) makes something a tiger (partly) explains both why tigers are striped and why being a tiger is explanatory.

I take it that Joseph LaPorte is suggesting just that in arguing as follows for *Fully Historical*:

> A lot is explained by an object's being a polar bear. That it is a polar bear explains why it raises cubs as it does, or why it has extremely dense fur, or why it swims long distances through icy water in search of ice floes.... The polar bear kind is a useful one for providing significant explanations. (2004: 19)

I take LaPorte to be claiming that the historical essence of polar bears (*Ursus maritimus*) explains how they came, in their environment, to have their phenotypic features. For objects to be polar bears, they must have a certain history and that history brought it about that these objects have the phenotype properties that they have. That's why being a polar bear is explanatory.

If this two-part argument for *Partly Historical Taxon Essentialism* is good, and the consensus were right in rejecting *Partly Intrinsic*, then we would have established that the essence of a taxon is *wholly* relational and historical: we would have established *Fully Historical*. But if this argument, based on the needs of historical explanations, is good, then how could the earlier argument, based on the needs of structural explanations, not be good? For, the present historical argument, of which we have just seen signs in the consensus literature, is analogous to the earlier intrinsic argument. And if that earlier argument is good, the consensus is not right in rejecting *Partly Intrinsic*: *Fully Historical* is false.

As noted in section 2.5, the following metaphysics underlies this sort of argument. Any organism in taxon *T* has certain phenotypic properties *because* it is in *T*; it is *because* an animal is an Indian rhino that it has one horn. So the property that *makes* something an Indian rhino, whether we call that property an "essence", "nature", or whatever, must *cause* that rhino, in its environment, to have one horn; the essence of Indian rhinos must explain the place that certain organisms have in the causal nexus *just because they are Indian rhinos*. The essence of *T* is whatever property, as a matter of actual fact, plays that fundamental causal role in an environment.[4] I have emphasized repeatedly that an essence must bear the explanatory burden; that is what I called the "Sober demand".

That concludes my discussion of question (A). I turn now to (B): What precisely is *Fully Historical Taxon Essentialism*? Okasha tells us that species are "particular chunks of the genealogical nexus" (2002: 200). But what exactly does this amount to? I shall start with two hypotheses I have found in the literature. But it is striking how little question (B) has been addressed effectively.

[4] We noted in section 2.9 that this view of essences generalizes to implements like cars.

It follows from our answer to question (A), that a *good* answer to question (B) must posit a taxon essence that can carry the burden of historical explanation. That is the crucial Sober demand. We shall see that the wholly relational essences posited by the two hypotheses do not meet the demand. To meet that demand, we need a partly intrinsic essence. *Fully Historical* should be abandoned.

3.3 Hypothesis (1): Descended from Certain Particular Actual Organisms

We seek an account of the historical essence of a taxon *T*, of the historical properties in virtue of which an organism is a member of *T*. Now, when we think of the history of, say, England or the wheel, we think of a series of particular events involving actual entities, one event leading to another. This leads naturally to the thought that the relevant history of a taxon *T* is a certain part of the history of all actual organisms in "the tree of life", one organism leading to another in particular reproductive events. Which part of the tree? Here's a suggestion. It's the part that includes the actual organisms that led to *T*: the taxon is identified by its historical origin *in those particular ancestors*, the organisms that constituted its *founder population*. Such an hypothesis certainly seems to be embraced by some philosophers of biology. Thus Mohan Matthen takes an organism to be a member of a particular species in virtue of standing "in a certain relation (*belonging to the same extended reproductive community*) as the originally ostended individual" (1998: 120). And consider the following passages:

> if we suppose that humans first appeared about a half million years ago, *Homo sapiens* is the name for the group that descended from the original organisms. (Ruse 1987: 344)

> we cannot produce any... *purely qualitative* specification of the essence of a kind like the tiger or the lion. However, if we are allowed to directly refer to particular individuals—e.g. a particular founding population—then we can provide such necessary and sufficient conditions (e.g. to be a tiger is to be descended from *this* ancestral population prior to any further speciation events occurring among the population's descendants). (Leslie 2013: 140, n. 24)

> to be of a certain species is to be [diachronically and developmentally continuous] with a stock of creatures from which the species actually evolved. (McGinn 1976: 135)

> it seems natural enough to take origins to be essential to taxa: you cannot possibly be a tiger unless you are part of the lineage that starts with the original tigers, and necessarily anything that is part of that lineage (without too many modifications) is a tiger. (Godman et al. 2020: 327)

What organisms go into the founder population of a species? In thinking about this we need to keep in mind that there is no precise moment at which one taxon evolves into another. At one time in a lineage there are organisms that are determinately *T1*s, at a later time there are organisms that are determinately *T2*s, and in between there are organisms that are not determinately *T1*s or *T2*s; as Hull says, "Obviously . . . there must have been instances in which non-horses (or borderline horses) gave rise to horses" (1978: 306). I take it that the founder population of *T2* must include, at least, those ancestors of *T2* that are the last determinate members of *T1* among *T2*'s ancestors. Thus, in a common situation, a geographic development isolates some members of *T1* which results, over time, in speciation and *T2*. The founder population of *T2* starts with those isolated members of *T1*. Perhaps it finishes with them? Or perhaps it includes all the intermediate organisms up to and including the first organisms that are determinately members of *T2*? We need take no stand on this.

The following hypothesis captures the core idea we are considering:

Hypothesis (1): The essence of *T* is its members' property of being descended from the particular actual organisms in *T*'s founder population.[5]

Is Hypothesis (1) complete? It nicely distinguishes zebras from horses: their founder populations are different. Still, where *T* is a species with a descendant species, more work has to be done to distinguish it from that descendant, for all the descendant's organisms are descended from *T*'s founder population too. What then marks out the *end* of *T*? Leslie's remark about "further speciation events" indicates the way to answer the question. We shall consider this in section 3.4. Meanwhile, we need not fuss about the incompleteness of Hypothesis (1) because it has many other problems.

First, suppose that the founder population for tigers includes a particular animal, Benji, perhaps a tiger. So, according to Hypothesis (1), there would have been no tigers if there had been no Benji. This is not plausible. And the hypothesis gets more implausible if Kripke's (1980: 112–15) "origin essentialism" is right, as

[5] Remarks like the following, arising out of the biological practice of introducing a new species term using a 'type specimen', suggested to me (2008: 362; 1.7 above) that some philosophers of biology might tie an essence to a type specimen:

> Any organism related to [the type specimen] in the appropriate ways belongs to its species, regardless of how aberrant the type specimen might turn out to be or how dissimilar other organisms may be. (Hull 1978: 311–12)

> Specify some individual, say Brigham Young, as your reference point, and then members of the same taxon are potential and actual interbreeders . . . (Ruse 1987: 344)

But such an essentialism cannot be right because many, presumably most, species don't have type specimens. So this hypothesis could not be a general account of a species' essence.

I argue it is (2018b; Chapter 4 below). For, according to Kripke, Benji would not have existed had his actual ancestors not existed. So the hypothesis makes the existence and reproductive role of every one of those ancestors, "back to the beginning", essential to the identity of *Panthera tigris*.

(a) Consider a species *S* that has an organism *O* as a member. As Alex Levine remarks, "having *O* as a member isn't constitutive of *S*'s identity" (2001: 333). Many agree:

> A species is not necessarily composed of those organisms which *happen to* compose it an any moment in its history. (Kitts and Kitts 1979: 615n)

> a *species* could exist even if many of the particular organisms that happen in fact to belong to that species did not belong to it . . . this . . . point seems as secure as any in philosophy. (LaPorte 2003: 584)

> Consider the set of organisms in *Homo sapiens*. I am one. If I did not exist, that set would not exist. Yet the species would. (Sober 1984: 337)

The point is that no one actual organism is essential to a species. And this includes any organism in the founder population (and hence in its ancestry). (b) Furthermore, *S*, including its founder population, could have had members that it did not actually have. Suppose, for example, that an organism, *O1*, that was a member of that founder population, had died pregnant. Had *O1* survived just long enough to give birth, there would have been another organism *O2* which might have been part of the founder population of *S*.

These objections to the hypothesis are intuitive but it is important to see that their support is stronger than intuition. Our working assumption is that the identification and naming of taxa in biology is successful in that it serves the purposes of historical and structural explanations (2.2, 3.1). Now suppose that objection (a) was not correct. So, for example, if Benji had not existed there would have been no tigers *even if his absence would have made no evolutionary difference*. All of the other members of the founder population would still have existed and led to a species indistinguishable from *Panthera tigris* but, according to Hypothesis (1), that species would not have been *Panthera tigris*. Suppose next that objection (b) was not correct. Then if the pregnant *O1* had survived and given birth to *O2* which was then part of the founder population of a species, that species would not have been *S* even if it was indistinguishable from *S*. There is surely no explanatory point to identifying species, or indeed any taxa, in this way.

In sum, the founder population of a species might not have had organisms that it actually had and it might have had organisms that it did not actually have.

That is the first problem for Hypothesis (1). Here's the second. An essence must be explanatory. I have argued that an essence of this historical and relational sort is "explanatorily hopeless" in that it could not carry the burden of *structural* explanations (2008: 362–3; 1.7 and 2.6 above).[6] Such explanations must show how the essence, along with the environment, cause a tiger to develop stripes. It is no help to be told that it is striped because it is descended from some Ur-tiger, Benji, and his associates. That does not tell us why any tiger, including perhaps Benji, is striped. So the essence cannot be *wholly* historical and relational in this way: we need an intrinsic component for structural explanations. But an analogous argument shows that such an historical essence could not carry the burden of *historical* explanations either. So this sort of historical property could not even be the historical *component* of the essence.

The historical explanation sketched in section 3.2 requires us to show that it is *because of* the historical essence of tigers, because of their history that they evolved to have stripes. That is the crucial Sober demand. Simply being in a population descended from certain animals could not alone explain this. What is the causal relevance to having stripes of simply being descended from Benji and company?[7] In the next section we shall explore the idea that the needed explanation must advert to the fact that tigers evolved from *a particular kind* of organism. That idea seems promising. But there is no promise in the idea that they must have evolved from *any particular organism* of that kind. If this is right, then we should not suppose that biologists, in their drive for explanation, have identified taxa that require descent from particular organisms rather than from particular kinds of organisms. Hypothesis (1) is not a good answer to our question (B).

3.4 Hypothesis (2): Descended from a Certain Kind of Organism (Which Itself has a Wholly Relational Essence)

The promising idea is that the historical essence of *T* is its having descended from a certain *kind* of organism not, as in Hypothesis (1), from certain ancestral *individuals*. Consider the following:

> "*Panthera tigris* = df the lineage descending from ancestral population P and terminating in speciation or extinction", P being... an appropriate population in the lineage that gave rise to today's tigers. (LaPorte 2004: 54)
>
> a domestic cat... is necessarily a member of the genealogical nexus between the speciation event in which the taxon originated and the speciation event at which it will cease to exist. (Griffiths 1999: 219)

[6] Also, "metaphysically" hopeless in contrast to its possible epistemic utility (2008: 362; 1.7 above).

[7] One might argue, and I do (2018b; Chapter 4 below), that Benji has a partly intrinsic individual essence that explains why he is striped (if he is). But how could this explain why Benji's descendants are striped? Only if that intrinsic property is essential to all tigers. But to suppose this is to abandon *Fully Historical Taxon Essentialism*.

> You and I are members of *Homo Sapiens*, therefore, because we both belong to the segment of the genealogical nexus which originated in Africa some 300,000 [years] ago (on current estimates), and which has not budded off any daughter species since that point. (Okasha 2002: 200–1)

Now perhaps these passages should be taken as proposing Hypothesis (1) but I think that they are more charitably understood as proposing that what is essential to a species is that it descended from a certain *kind* of organism rather than from certain individuals. The kind is of those organisms in the prior species that partly, perhaps completely, constitute the founder population.

In thinking about this proposal, we need to keep in mind that, according to the consensus view that we are trying to elucidate, the essences of all taxa are *wholly* relational: *Partly Intrinsic Taxon Essentialism* is rejected. So this relational view applies not only to the domestic cat but also to its ancestor species, to the ancestor of that species, and so on back to the beginning of life. Nothing intrinsic about any taxon in that tree of life, or any part of it forming a founder population, is essential; "it's relations all the way down". So the core proposal is the following hypothesis:

> **Hypothesis (2):** The essence of *T* is its members' property of being descended from members of a certain different taxon T* (where T* itself has a similar wholly relational essence).

Like Hypothesis (1), Hypothesis (2) needs an addition to mark out the end of a species that has a descendant species, and hence to move (2) toward completeness. Set that aside for a moment. Hypothesis (2) fails the completeness test on other grounds as I have, in effect, pointed out in "Resurrecting":

> suppose that species *A* splits into species *B* and *C*, then *B* splits into *D* and *E* and *C* splits into *F* and *G*.... What does [Hypothesis (2)] tell us about the nature of *B*? That *B* is descended from *A* and that *B* is distinct from *C*. And that's what [Hypothesis (2)] tells us... about the nature of *C* too. So that clearly does not distinguish *B* from *C*. No more does it distinguish *D* and *E* to know that each is descended from *B*, and *F* and *G* to know that each is descended from *C*. Furthermore, since [Hypothesis (2)] does not distinguish *B* from *C* it does nothing to distinguish *D* and *E*, descended from *B*, from *F* and *G*, descended from *C*. (2008: 367–8; 1.9 above)

In brief, the hypothesis fails to distinguish zebras from horses.

Related to this incompleteness, Hypothesis (2) is another hypothesis that is explanatorily hopeless. It provides nothing that could explain the phenotypic differences between species, nothing that could meet the crucial Sober demand.

Partly because of the way a taxon is, partly because of its essence, it has *evolved* to have its phenotypic properties, and each of its members *develops* those properties (in its "normal" environment). Historical explanations of those evolutionary facts, and structural explanations of those developmental facts, must advert to that essence. But, on Hypothesis (2)'s wholly relational view, there is nothing in the essence that could carry this explanatory burden. Why did tigers evolve to have stripes? It is no answer to say that they descended from *T**, which has no nature beyond having descended from *T***, which has no nature beyond having descended from *T****, and so on. There is nothing in such purely relational natures that has any causal relevance to having stripes or to having any other phenotypic property. Similarly, what explains the development of a striped mature tiger from this unstriped zygote, of a one-horned rhino from that hornless zygote, and so on? A purely relational nature cannot explain that because it does not advert to anything that could cause stripes or horns.

Kevin de Queiroz (1992, 1995) has a suggestion that may appear to add more substance to the historical essence of *T*, but the appearance is an illusion. His proposal for the essence of *Mammalia* is: "it is logically necessary for an organism to be part of the clade stemming from the most recent common ancestor of horses and echidnas to be a mammal" (1995: 224.). LaPorte develops this idea, taking account of the fact that there could have been mammals even if there had not been any horses or echidnas. He gives the name '*G*' to "the group that happens, as a matter of contingent fact, to be the most recent ancestor common to both the horse and echidna". The essence of *Mammalia* is then to be "in *G*, or descended from *G*" (2004: 12). The extra substance may seem to come from anchoring *Mammalia* in actual taxa, the horse and the echidna. But the extra depends on the essences of the horse and the echidna. Can the pattern of explanation for *Mammalia* be extended to them? Perhaps, but not much further. Thus, echidnas are a family, *Tachyglossidae*, made up of three genera. So, the pattern could explain the essence of echidnas in terms of its three genera, including the long-beaked echidna, *Zaglossus*; and then we could explain each genus in terms of its species; thus, *Zaglossus* could be explained in terms of its species, including the Eastern long-beaked echidna, *Zaglossus bartoni*. But what then? The pattern of explanation for *Mammalia* has no answer. So, all the explanations that *rest on* such an answer fail. We need a new approach to explain the essence of *Zaglossus bartoni*. And if we had that approach, de Queiroz's proposal would be redundant.

The demonstrated incompleteness and explanatory inadequacy of Hypothesis (2) are the decisive objections to it. But the hypothesis also has the other completeness problem we mentioned. What addition could be made to mark out effectively the end of a species (with a descendant species)? Suppose that there is a group of organisms *G1* that are all the members of a species *S* at a certain time *t* and *G2* is a group of immediate descendants of *G1* (though not necessarily *all* of its immediate descendants). Now if *G2*'s members are also members of *S* then *S* did not end at *t*. So what has to be the case for *G2*'s members *not* to be members of *S* and hence for *S* to end? Clearly, as some of the quoted passages note, there has to be a *speciation event* between *G1* and *G2*. But then what constitutes a speciation event? I shall explore this vexed question briefly, as it bears on *Fully Historical Taxon Essentialism*.

An appealing contribution to an answer, given by *Partly Intrinsic*, is that a speciation event is a change in intrinsic natures: the members of *G2* are *intrinsically different* from the members of *G1*. But *Fully Historical* must reject that, giving a wholly relational answer.

A popular relational answer, stemming from Willi Hennig (1966: 207–11) and reflecting the influence of cladism, is that a speciation is *a split of an old species into two new species*. The members of *G2* are not in the same species as the members of *G1* in virtue of *G1* having another group of immediate descendants, *G3*, that is a distinct species from *G2*. In virtue of what are *G2* and *G3* distinct species? We look to a "species concepts" for an answer: perhaps *G2* and *G3* are reproductively isolated; perhaps they occupy different niches.

So the view is that a split of this sort is both necessary and sufficient for a species (with a descendant species) to end. And this seems to be the only somewhat plausible way for *Fully Historical* to explain the end of a species in relational terms. Yet it is not that plausible. And it is certainly "contentious... in evolutionary theory" (Pedroso 2012: 186).

First, the claim that a split is necessary for speciation rules out anagenesis. This is worryingly extreme, for a reason given earlier (1.9). Suppose that *Homo sapiens* had evolved from protists without any splits. Then, given the rejection of anagenesis, all the organisms in this lineage would be in the same species. Kitcher aptly notes that "this strikes many people as counterintuitive (even insane)" (2003: 151). More importantly, the rejection seems at odds with actual taxonomic practice. Taxonomists seem quite ready to contemplate anagenesis (e.g. Kimbel et al., 2006; Strotz and Allen, 2013; Heaton, 2016).

So it seems doubtful that a split is necessary for a species with descendants to go extinct. But, looking at actual taxonomic practices again, it seems doubtful that it is sufficient either. Thus Sterelny remarks: "Some, perhaps most, evolutionary biologists take speciation to occur only when there have been intrinsic changes" (1999: 130). (Sterelny is puzzled by this, given the consensus on *Fully Historical*, conjecturing that biologists are influenced by the folk!) And what about the practice of non-evolutionary biologists?[8]

We saw earlier that Hypothesis (2), the most promising wholly relational view of a taxon's essence, fails to distinguish species and, relatedly, is explanatorily hopeless. Furthermore, the hypothesis needs an addition to explain how a species

[8] On these practices, see Dupré (2002: 82).

(with a descendant species) can end. We have just seen that the popular view that a split is necessary and sufficient for this end is dubious at best.

The signs are that there is no good answer to our question (B): that there is no plausible wholly relational essence that could carry the burden of historical explanation. It looks as if we should abandon *Fully Historical Taxon Essentialism*.

3.5 Hypothesis (3): Descended from a Certain Kind of Organism with a Partly Intrinsic Essence

The problem with Hypothesis (2) is that an historical component to a taxon's essence *requires* an intrinsic component; it needs that component to make the hypothesis complete and plausible. As John Dupré, who rejects essentialism altogether, has aptly remarked,

> the phylogenetic criterion must be parasitic on some other, synchronic, principle of taxonomy. (1981: 89)
>
> being descended from one of the members of a particular set is no criterion at all unless there is some way of picking out the members of the set. (1993: 57)

We must go against the consensus by combining *Partly Historical* with *Partly Intrinsic*.

The promising idea underlying Hypothesis (2) was that the relevant history of taxon *T* is of organisms of one kind evolving into organisms of another kind, until we finally reach *T*. We need to incorporate into this the view that these kinds have partly intrinsic essences:

> **Hypothesis (3):** The historical component of the essence of *T* is its members' property of being descended from members of a certain different taxon *T**, where *T** has an intrinsic underlying component to its essence.

What about the end of *T*? If we allow anagenesis, then that end is marked out solely by the intrinsic component of the essence and so we need say nothing about it in characterizing the historical component. If, on the other hand, and implausibly, anagenesis is rejected, then Hypothesis (3) would require an addition capturing that the end of a species that has descendant species is marked out not only intrinsically but also by a split. I shall continue without the addition.

Why believe Hypothesis (3)? Why do we have to suppose that *T**, from which *T* evolved, has a partly intrinsic essence? Because the essence of *T** must feature in the historical explanation of how *T* evolved, via its founder population, to have the properties it has. We have seen that the mere fact that *T* evolved from *T**, which in turn evolved from *T***, and so on, does not provide any "substance" that could carry the explanatory burden of *T*'s essence. It could not meet the crucial Sober demand. We meet that demand with intrinsic properties.

Consider some examples, starting with the polar bear. As LaPorte points out, the bear's historical essence is supposed to explain how it came to have extremely dense fur (3.2). That it evolved from *T** will only explain that if there is something that the bear's founder population inherited from *T** that is central to the explanation of its having that fur. That something has to have played a crucial role in *causing* the bear's fur. And that something *has to be something intrinsic.* Think next of Stephen Jay Gould's lovely story, "The Panda's Thumb" (1981), about another *Carnivora*, the Giant Panda (*Ailuropeda melanoleuca*): the panda's ancestor had a sesamoid bone that evolved into the thumb of the panda. So, a crucial feature of the founder population is the part of its intrinsic underlying nature that produced its sesamoid bone and that the panda inherited. The historical explanation of the thumb demands a history that includes an ancestor with that essential feature. Finally, suppose that we want to explain how taxon *T* came to have eyes. A model has been proposed of a series of minor modifications taking us from a taxon with a simple light-sensitive patch to a taxon with a focused lens eye (Nilsson and Pelger 1994). As Brett Calcott points out: "this explanation makes an implicit bet: that there is genotypic variability available that maps smoothly to these small morphological changes" (2009: 59). This variability is in the intrinsic underlying natures of the taxa in this series. Facts about those underlying natures are fundamental to the explanation of the evolution of eyes. The historical component of the essence of *T* includes having a series of ancestors with those intrinsic natures.

According to Hypothesis (3), an organism is a member of *T* in virtue of having a certain intrinsic underlying property, *P*, *as a result of a history of the following sort*: *T* evolved from *T** with a certain different intrinsic underlying property, *Q*, which evolved from *T*** with a certain other different intrinsic underlying property, *R*, which evolved... and so on back to the beginning of the tree of life.

Hypothesis (3) provides an account of *Partly Historical*'s historical component, an account that presupposes *Partly Intrinsic*'s intrinsic component. This yields an account of a taxon's essence that is *complete* in that it distinguishes one taxon from another and plausible in that it can carry the burden of structural and historical explanations. And Hypothesis (3) entails that the wholly relational view urged by the consensus, *Fully Historical*, is false.

There are, of course, well-known objections to any sort of intrinsic biological essentialism. I have responded to these in Chapters 1 and 2. If, nonetheless, these objections were effective, then I think we would still be left with no complete and plausible account of *Fully Historical*.

3.6 Objections: Twin Earth and the Like

Here are three likely objections from the philosophical consensus.

Twin-Earth-Tigers: Suppose that there was a Twin Earth with a qualitatively identical tree of life to that on Earth. So, there would be a species on Twin Earth that would be intrinsically identical to our tigers and would have evolved from ancestors and a founder population that were intrinsically identical to the ancestors of our tigers. So, according to Hypothesis (3), these Twin-Earth-tigers would have the same essence as tigers. Yet, the consensus objection runs, they would not *be* tigers. So Hypothesis (3) is not complete after all.

Engineered-Tigers:[9] Here is a much more empirically plausible possibility. Suppose that we genetically engineered new organisms "from scratch", organisms bearing no lineal relationship to existing tigers yet which are genetically and phenotypically indistinguishable from those existing tigers and capable of interbreeding with them to produce fertile offspring. According to Hypothesis (3), indeed according to (1), (2), or any likely biological hypothesis, these engineered-tigers would not be tigers. Yet surely they are. So these hypotheses are wrong.

Semi-Engineered-Tigers: Not only *can* these engineered-tigers interbreed with existing tigers, many of them *do*. The result is a population made of three groups, indistinguishable except by their origins: (a) descendants solely of historical tigers; (b) descendants solely of engineered-tigers; (c) descendants of both, "semi-engineered-tigers". According to our hypotheses, not only are (b) not tigers, (c) are not. Yet surely they are. So these hypotheses are wrong.

Philosophers are full of intuitions about the possibilities of Twin Earth and the like. Yet, from the naturalistic perspective, we should resist the temptation of relying on them (Devitt 2006). Rather, we should look to science to respond to these possibilities. With the present objections, we should look to the demands of biological explanation. For the essences of taxa must meet the demands of these explanations.

Before responding to these objections, it is interesting to consider this question: "Did the taxa posited by pre-Darwinian biologists have partly historical essences?" It seems likely that they did not. Prior to the theory of evolution, biologists, when positing taxa, were not concerned with historical explanations like the ones we have illustrated. Of course, those biologists had the same interest in structural explanations as current biologists. So, the taxon picked out by their use of 'polar bear' had the intrinsic component of the essence of the taxon picked out by the current use, but it seems implausible to suppose that it had the historical

[9] I owe this objection to an anonymous referee.

component. So, though the two uses of 'polar bear' doubtless have the same extension, they do not have the same intension.[10]

I start with the easiest objection, engineered-tigers. As the history of discussions of "Swampman" show, many philosophers will have the firm intuition that engineered-tigers are tigers. But, from the naturalistic perspective, this intuition does not hold up (Neander 1996). The tigers posited by post-Darwinian biologists are animals with a history that biologists are very much interested in explaining. So, biologists posited a kind with an essence that is central to historical explanations of phenotypic properties. Engineered-tigers lack that essence. So, contrary to those firm intuitions, engineered-tigers are not tigers. The structural explanations for tigers will work just as well for engineered-tigers, of course, but not the historical ones.

The same goes for semi-engineered-tigers: they are not tigers. Still, should that group expand to the point that its size becomes large relative to tigers (the descendants solely of historical tigers), one supposes that there would be pressure on biologists to include them in the population that they theorize about. One can imagine a conceptual change: biologists stop talking of "tigers" and start talking of "new-tigers", a population made up of historical tigers *and* semi-engineered-tigers, perhaps even of engineered-tigers also. Biologists would, of course, be as interested in the historical explanation of the phenotypic properties of new-tigers as they are of our tigers, but the historical component of the essence that features in those explanation of new-tigers would take account not only of evolution but of engineering.

Finally, we must consider Twin-Earth-tigers. Biologists would have the same interest in historical explanations of Twin-Earth-tigers as of our earthly tigers, and their shared essences, according to Hypothesis (3), would play the same role in the explanations of each sort of animal. According to Hypothesis (3), Twin-Earth-tigers would be tigers, just as the objection claims. But perhaps Hypothesis (3) is indeed incomplete. Perhaps the essence of our tigers is implicitly tied to the location Earth and so Twin-Earth-tigers are not tigers? Or, more precisely, since Twin Earth's tree of life could occur here along with our actual tree, perhaps the essence of tigers is implicitly tied to the *particular primitive organisms that actually began life on Earth* (even though particular organisms are inessential from then on; section 3.3)? I can see no fact of the matter about this. Our present taxonomic practices are simply not geared to yield a determinate answer in light of the bizarre possibility of Twin-Earth-tigers. So, it is indeterminate whether or not Hypothesis (3) needs to be enhanced by tying essences to the particular beginning of life on Earth and hence whether or not Twin-Earth-tigers are tigers. But suppose I'm wrong about this indeterminacy and Hypothesis (3) does need to

[10] I owe this paragraph to my former student Raj Nanavati.

be thus enhanced to allow for this philosophical fantasy. That seems of little theoretical interest.

3.7 Conclusion

The consensus view is *Fully Historical Taxon Essentialism*: the essence of a taxon, particularly a species, is wholly relational, particularly historical. This raised two question: (A) Why believe this doctrine? (B) What precisely is this essence?

In addressing (A), I argued that the historical explanation of the phenotypic properties of a taxon demand an essence that has an historical component (3.2). So, this is an argument for *Partly Historical Taxon Essentialism*. The argument drew on the literature and is analogous to one I have given earlier (2008 and Chapter1 above) for the doctrine, rejected by the consensus, that the essence has an intrinsic component. If the consensus were right to reject that doctrine, *Partly Intrinsic*, then the argument from historical explanation would be an argument for *Fully Historical*. But we should wonder how that argument could be good and yet the analogous one for *Partly Intrinsic* not be.

An answer to (B) must be complete in that it distinguishes one taxon from another; for example, zebras from horses. And it must be plausible in that it meets the crucial Sober demand: the essence must bear the burden of historical explanations of phenotypic properties.

Hypothesis (1) is certainly embraced by some philosophers of biology. It is the view that the essence of *T* is its members' property of being descended from the particular actual organisms in *T*'s founder population (3.3). But *T* might have evolved from a founder population made up of different particular organisms: that population could have lacked any of the organisms it had and might have had organisms it did not have. Furthermore, the hypothesis failed to satisfy the Sober demand on explanation.

Hypothesis (2) is a more promising hypothesis, signs of which are to be found in the literature: the essence of *T* is its members' property of being descended from members of a certain different taxon T* (where T* itself has a similar wholly relational essence) (3.4). But this hypothesis did not pass the completeness test: it fails to distinguish one species from another. Relatedly, it is explanatorily hopeless. Why did tigers evolve to have stripes? It is no answer to say that they descended from *T**, which has no nature beyond having descended from *T***, which has no nature beyond having descended from *T****, and so on. Furthermore, the hypothesis needs an addition to explain how a species (with a descendant species) can end. The popular view that a split is necessary and sufficient for this end is dubious at best.

I concluded that there is no good answer to our question (B) and we should abandon *Fully Historical Taxon Essentialism*. Our answer to (A) gives us reason to believe in an essence that has an historical component, but it also has an intrinsic one. We should adopt Hypothesis (3): the historical component of the essence of *T* is its members' property of being descended from members of a certain different taxon *T**, where *T** has an intrinsic underlying component to its essence. The history that constitutes the essence of *T* is of organisms of one intrinsic underlying kind evolving into organisms of another intrinsic underlying kind, via founder populations, until we finally reach *T* (3.5). Hypothesis (3) survives objections from Twin Earth and the like by focusing on the fact that the essences of taxa must meet the demands of biological explanations (3.6).

Combining *Partly Intrinsic Taxon Essentialism* with this account of *Partly Historical Taxon Essentialism* yields the following doctrine (for taxa thought to fall under the Linnaean hierarchy):

> *Taxon Essentialism*: The essence of a biological taxon is its members' property of having a certain intrinsic underlying property as a result of descent from members of a certain different taxon, where that taxon has a certain different intrinsic underlying component to its essence.

We move now to consider the essence of individual organisms.

4
Individual Essentialism in Biology

4.1 Introduction

Consider the taxa that are thought to fall under the biological categories in the Linnaean hierarchy: kingdoms, phyla, classes, orders, families, genera, species, and even subspecies (varieties).[1] Saul Kripke (1980) and other metaphysicians have proposed, on the basis of modal intuitions, two distinct essentialist doctrines about these taxa. The first, *Partly Intrinsic Taxon Essentialism*, is the doctrine that these taxa, particularly species, have essences that are at least partly intrinsic underlying, and probably largely genetic, properties. As we have seen in Chapters 1 and 2, this has long been the subject of criticism in biology and the philosophy of biology. Contrast this treatment with that of the second doctrine:

> *Essential Membership*: If an individual organism belongs to a taxon it does so essentially.

Until Joseph LaPorte's "Essential Membership" (1997), this doctrine had, as he remarks, "largely escaped the attention of philosophers of science". Whereas, he charges, "essentialists have tended to be rather naïve on scientific matters", he aims to approach the issue "in the light of biological systematics" (p. 97). This approach leads him to reject *Essential Membership*. Samir Okasha (2002) endorses LaPorte's rejection.[2] Thus, LaPorte and Okasha both urge, *from a biological basis*, a view of what is *not* essential to an individual organism. But neither they nor, so far as I can discover, any other philosopher of biology or any biologist, have seriously addressed the issue, broader than *Essential Membership*, of what *is* essential to the organism. It seems that this issue, much discussed by metaphysicians (e.g. McGinn 1976; Salmon 1979; Kripke 1980; Forbes 1986; Robertson 1998; Mackie 2006), has *entirely* escaped the attention of philosophers of biology. I propose to give it attention.

Doing so raises a worry like that of Kingsley Amis' "Lucky Jim". Jim is contemplating his opening to a paper that he is desperately trying to publish.

[1] This chapter modifies and expands the discussion in Devitt (2018b).

[2] I infer from the form of LaPorte's argument that he thinks that philosophers of biology up to that time had *implicitly* rejected *Essential Membership*. Since then, Griffiths (1999: 219), Levine (2001), and Leslie (2013: 139) have done so explicitly.

Biological Essentialism. Michael Devitt, Oxford University Press. © Michael Devitt 2023.
DOI: 10.1093/oso/9780198840282.003.0004

The paper is on "The Economic Influence of Developments in Shipbuilding Techniques 1450–1485". His opening is: "In considering this strangely neglected topic...". Jim's worry is that the topic may be neglected *for good reason*. We should worry that the metaphysical topic of individual biological essentialism may also be neglected by philosophers of biology *for good reason*. Perhaps it is of no scientific interest. I shall attempt to show that it is of interest.

In "Resurrecting Biological Essentialism" (2008), and in Chapters 1 and 2 above, I argued, *against* the consensus, for *Partly Intrinsic Taxon Essentialism*:[3] there is an intrinsic component to the essence of a taxon *T*, to "*Et*". In "Historical Biological Essentialism" (2018a) and Chapter 3 above, I argued, *with* the consensus, for *Partly Historical Taxon Essentialism*:[4] there is an historical component to *Et*. So, for an organism *O* to be a member of *T* it must have both components.

Those discussions address the issue of *taxon* essentialism. The present chapter addresses the issue of *individual* essentialism. What it is to be *O*? What is the "essence", "nature", or "identity" of *that very individual organism*? What is the property *in virtue of which* an object is *O*, what *constitutes* its being *O*, what *makes* it *O*? I shall use analogues of the arguments in the taxon discussions to argue that *the demands of biological explanation* support the Kripkean view that *O* has an essence, "*Eo*", that is *also* partly intrinsic and partly historical:

> *Partly Intrinsic Individual Essentialism*: Biological individuals have essences that are partly intrinsic underlying, probably largely genetic, properties.[5]
>
> *Partly Historical Individual Essentialism*: Biological individuals have essences that are partly historical properties.[6]

So my essentialism is based on biology, not simply on intuition. Armed with a view of *Et* and *Eo*, it is easy to assess *Essential Membership*: *O* is essentially a member of *T* iff its having *Eo* entails its having *Et*. I shall present a case for *Essential Membership*. Finally, I will respond to LaPorte's objections to *Essential Membership*. One these objections is of considerable *general* interest.

I shall argue for the Kripkean view of *Eo* in sections 4.2–4.5. On the basis of these discussions I shall present the case for *Essential Membership* in section 4.6. Finally, in sections 4.7 to 4.10, I shall respond to LaPorte's objections.

Two points about the argument for *Essential Membership* in section 4.6. (1) Suppose that both *Partly Historical Taxon* and *Partly Intrinsic Taxon* are right. Then clearly for *Essential Membership* to be right also it has to be the case that, for any *O* that is a member of *T*, *Eo* must include both the historical and intrinsic components of *Et*. And clearly, if the controversial *Partly Intrinsic Taxon* is not

[3] The doctrine was called "Intrinsic Biological Essentialism" in my 2008 (also in my 2020, 2021a).
[4] The doctrine was called "Partly Historical Essentialism" in my 2018a.
[5] This doctrine was called "Intrinsic Individual Essentialism" in my 2018b.
[6] This doctrine was called "Historical Individual Essentialism" in my 2018b.

right, and only the relatively uncontroversial *Partly Historical Taxon* is right, then *Essential Membership* requires only that *Eo* includes the historical component of *Et*.

(2) The other point concerns *Partly Historical Taxon* and is more tricky. Despite the fact that the doctrine is part of the consensus in biology, we have seen in Chapter 3 how little has been said about what precisely the historical essence of a taxon is. The nature of this essence is something else that has largely escaped attention. And, I argued, the little that has been said is not plausible in that the proposed essence cannot carry the explanatory burden; it does not meet the "Sober demand". I urged the view that the relevant history is of organisms of *a certain intrinsic kind* evolving into organisms of *a certain other intrinsic kind*, until we reach the taxon in question. This provides, of course, another argument for the controversial *Partly Intrinsic Taxon*. So it is sure to be controversial too. So my argument for *Essential Membership* on the basis of this view will be controversial. But what alternative view of the essential history would the consensus prefer? The best I can do is to *predict* that on any plausible account of the historical component of *Et*, *Essential Membership* will hold.

Finally, *Essential Membership* has become topical because of a series of papers mostly in *Biology and Philosophy*, beginning with the one by Alex Levine (2001). Levine rejects *Essential Membership* and so holds that any organism is only contingently a member of its species. He finds this contingency in conflict with the common thesis in biology that any organism selected as the "type specimen" for a species is necessarily a member of that species. Levine expresses the conflict neatly: "*qua organism*, the type specimen belongs to its respective species contingently, while *qua type specimen*, it belongs necessarily" (p. 334). LaPorte (2003), Matthew Haber (2012), Joeri Witteveen (2015), and Jerzy Brzozowski (2020) have struggled with this alleged conflict. The truth of *Essential Membership* would remove the apparent conflict. So too, of course, would the falsity of the thesis about type specimens. I argue in Chapter 5 below that this thesis is indeed false. (Does this falsity yield another conflict, this time with the *truth* of *Essential Membership*? We shall see that it does not in section 5.3.)

I turn now to the issue of individual essentialism.

4.2 Individual Essentialism: A Kripkean View

What is *Eo*, the essence of an individual biological organism *O*? I started this chapter by noting that this question has entirely escaped attention in the philosophy of biology. Even LaPorte (1997), Okasha (2002), and Levine (2001), do not really address this question in their rejection of *Essential Membership*. Because they go along with the consensus view that *Et*, the essence of a taxon *T*, is wholly historical, their argument against *Essential Membership* requires them to argue that *Eo does not* include the relational properties that constitute *Et*. But the argument does not require them to take a stand on what *Eo does* include and they do not do so. So they have nothing to say on what properties, relational and/or intrinsic, *do* constitute *Eo*. Nor do they cite any other philosophers of biology, or any biologists, as having something to say on the matter,[7] and I know of none who have. Yet, as a result of Kripke's answer to the question—his so-called "origin essentialism"—the question has received much attention from metaphysicians, as we shall see. I shall start my discussion with Kripke's answer.

First, we must distinguish the sort of essentialist question that concerns us from another that Kripke describes as follows: "What properties must an object retain if it is not to cease to exist, and what properties of the object can change while the object endures?" (1980: 114, n. 57). In the present context, this is the issue of whether an individual organism "could cease to belong" to a certain biological taxon (LaPorte 1997: 98). Kripke is not concerned with this sort of "temporal question" but rather with the following sort: "What (timeless) properties could the object not have failed to have, and what properties could it have lacked while still (timelessly) existing?" (1980: 114, n. 57). In the present context, this is the issue of whether an individual organism "could have failed, from its inception" to belong to a certain biological taxon (LaPorte 1997: 98).

Considering one particular organism, Kripke famously asked: "could the Queen... have been born of different parents from the parents from whom she actually came?" Taking her parents to be "the people whose body tissues are sources of the biological sperm and egg" that led to the Queen (1980: 112), Kripke answers in the negative: "anything coming from a different origin would not be this object" (p. 113). He moves on to raise a similar question about a particular table: "could *this table* been made from a completely *different* block of wood, or even of water cleverly hardened into ice...?" (p. 113). Once again he answers in the negative (p. 114). So he is urging "the principle that the *origin* of an object is essential to it" (p. 114, n. 57). What about the principle "that the *substance* of which it is made is essential"? Kripke accepts this principle too: the wooden table in question could not "have been made of anything other than wood" (pp. 114–15, n. 57). Finally, Kripke notes a relationship between the two principles. Supposing, as Kripke does, that it is essential to the block from which the table was made that it was a block of *wood*, then it follows from the origin principle that it is essential to the table that it is made of wood (p. 115, n. 57).[8]

According to the origin principle it is essential to the Queen that she came from certain parents and from certain gametes. Those gametes united to form a zygote. According to the origin principle it is also essential that the Queen came from that

[7] Nor do the papers that responded to Levine: LaPorte (2003), Haber (2012), Witteveen (2015), and Brzozowski (2020).

[8] Kripke claims also that "(roughly) *being a table* seems to be an essential property of the table" (p. 115, n. 57). I have argued that this cautious suggestion is mistaken (2005: 156).

particular zygote. What we have said about the Queen applies, of course, to her parents, their parents, and so on. So the historical component of *Eo* is constituted by the actual individual people, gametes, and zygotes that make up that ancestral history, that family tree.[9]

Kripke does not apply the substance principle to the Queen but we can do so. Just as the table must come from a block that is essentially wooden material, the Queen must come from a zygote that is essentially human material. But, in the spirit if not the letter of Kripke, we should surely go further with the substance of the Queen: what is essential to her zygote, hence to the Queen herself, is not simply that the zygote is constituted of human material of *some sort* but of *the particular sort in that zygote*. So, we have the Kripkean proposal that *Eo* is partly *O*'s origin in a certain zygote formed from certain gametes, hence from certain parents, and *Eo* is partly intrinsic properties of *O*'s zygote.[10] So, on this Kripkean proposal, *Eo*, like *Et*, has both an historical component—*Partly Historical Individual Essentialism*—and an intrinsic one—*Partly Intrinsic Individual Essentialism*.

The support for these Kripkean views about individual essences has so far come from intuitions. This raises an important methodological question. Okasha claims that it is "widely held" that "claims about individual essence", unlike those about "kind essence", "are not responsible to empirical science"; they are matters "for the armchair metaphysician" to be handled by "consulting... modal intuitions" (2002: 193). If this were so, it would be bad news for these claims. But it is not so.

It would be bad news, first of all, because intuitions may not be shared. We shall see that those supporting *Partly Historical Individual* are not shared by some influential philosophers. Second, and more important, intuitions about what is essential, like intuitions about any area of reality, are empirical judgments that provide at best indirect evidence about the topic; or so I have argued (2006). These judgments need the support of more direct evidence. So it is fortunate that the Kripkean intuitions can be given empirical support, as I shall now show.

4.3 An Argument for *Partly Intrinsic Individual Essentialism*

Let us start with an argument for *Partly Intrinsic Individual*. In section 4.1, I foreshadowed that I would argue for my positions on *Eo* using analogues of earlier arguments for positions on *Et*. The earlier argument I shall be drawing on

[9] Penelope Mackie raises the question: "Are we to suppose that a non-actual possible world contains the Queen only if it branches off the actual world at some time *after* the Queen actually came into existence?" (2006: 96). And the answer is "No": the non-actual possible world can branch off from the actual world at any time but it must always contain the Queen's actual ancestral history of those actual entities.

[10] This is only true of sexual organisms, of course.

in this section is a two-part one for *Partly Intrinsic Taxon Essentialism* (2008: 351–5; 1.3, 2.3 above).

In the first part, I argued that biological generalizations about the phenotypic properties—morphology, physiology, and behavior—of species and other taxa have explanations that advert to intrinsic components of essences. The explanations that featured in the argument were "structural" ones about the underlying developmental mechanisms in members of a taxon that make the generalizations true. As Kitcher notes, "explanations of this type abound in biology: think of the mechanical accounts of normal (and abnormal) meiosis, of respiration and digestion, of details of physiological functioning in all kinds of plants and animals" (1984: 121). In contrast, "historical" explanations tell us how members of the taxon came to have such mechanisms. Those explanations feature in my argument for *Partly Historical Taxon* (2018a; Chapter 3 above), an analogue of which will be used in the next section to argue for *Partly Historical Individual.*

In the second part, and relatedly, I argued that a taxon's intrinsic essence explains *why* being in the taxon is explanatory.

So here is the first part of my analogous two-part argument for an intrinsic component to the essence of an *individual organism*, *O*, for *Partly Intrinsic Individual.* Structural explanations of the phenotypic properties of *O* advert to the intrinsic component of the *O*'s essence. The properties of *O* that I have in mind are what *O* looks like, what it eats, where it lives, what it preys on and is prey to, its signals, its mating habits, and so on. Claims about these properties of *O* may not be *as* biologically interesting as generalizations about the members of a taxon but they are interesting nonetheless. Indeed, the generalizations rest on them. Thus, biologists claim not only that ivy plants grow toward the sunlight (Sober 1993: 6) but that a particular organism *O1* does; they claim not only that polar bears have white fur (p. 21) but that *O2* does; they claim not only that Indian rhinoceri have one horn and African rhinoceri have two (p. 21) but that *O3* has one horn and *O4* has two; they claim not only that the Australasian bittern is superbly camouflaged (Sterelny and Griffiths 1999: 32), but that *O5* is; they claim not only that "Australian trees... are not just drought-proof; they are fireproof as well" (p. 203) but that *O6* is.

Claims of this kind demand an explanation. *Why are they so*? Set aside historical explanations of what *led to* there being those organisms with those properties and seek structural explanations. We need an account of *what it is about the organisms that made* the claims true. What are the developmental mechanisms? The truth of these claims cannot be brute facts about the world and so must be explained. Explanations will make some appeal to the environment but they cannot appeal only to that. There has to be something about the very nature of an organism that, given its environment, determined the truth of the claim. That something is an intrinsic underlying, probably largely genetic, property, details to be discovered by biologists. Some intrinsic underlying property of the Indian rhino O3 caused it, in its environment, to grow just one horn. A different such property of the African rhino O4 caused it, in its environment, to grow two horns. The intrinsic difference explains the phenotypic difference. If we put together each intrinsic underlying property that similarly explains a similar claim about an organism, then we have the intrinsic component of its essence.

Note that the structural explanations on which the argument for *Partly Intrinsic Individual* rests, like the earlier Kitcher ones about meiosis, etc. on which the argument for *Partly Intrinsic Taxon* rests, are biological ones not "folk" ones. And, I emphasize, these arguments look to biologists to discover precisely what intrinsic underlying properties do explain the phenotypical properties and hence are essential. That is only appropriate.

I turn next to an analogue of the second part of the earlier argument for an intrinsic component to a *taxon's* essence. Just as the generalizations about taxa reflect the fact that it is *informative* to know that an organism is a member of a certain taxon (Sterelny and Griffiths 1999: 195), so too do the claims about individual organisms reflect the fact that it is *informative* to know the identity of an organism. And just as being a member of a certain taxon is more than informative, it is *explanatory* (Matthen 1998: 115; Ereshefsky and Matthen 2005: 2–3), so too is being a certain organism. The fact that *O* is a tiger, an Indian rhino, an ivy plant, or whatever, explains a whole lot about its phenotypic properties; biological classifications are explanatory, as noted earlier (2.2). *Why are they*? Because the essential nature of a taxon, to be discovered by biologists, causes its members, in their environment, to have those phenotypic properties. Now, there may well be no sign that biologists are similarly committed to the idea that *O* is striped because it is a certain organism, say the tiger Benji. But, by parity of reasoning, they should be. The fact that *O* is Benji explains a whole lot about *O*'s phenotypic properties, including many, like a particular face marking, that *O* has *not* simply because it is a tiger. *How does that fact explain*? Because the essential nature of Benji, to be discovered by biologists, causes him, in his environment, to have those properties. What nature? If our concern is structural, so it is with a nature that causes Benji's development into an organism with those properties, the nature must be intrinsic. So the same intrinsic nature or essence that (partly) makes something Benji (partly) explains both why he is striped and why being Benji is explanatory. Sober rightly insists that the essence of a species must explain why its members are the way they are. It must be "a causal mechanism that acts on each member of the species, making it the kind of thing that it is"; the essence must explain why the members "are the way they are" (1980: 250). There is the same "Sober demand" on the essence of Benji. The intrinsic essence of Benji is a causal mechanism that makes him the thing that he is, that explains the way he is.

In sum, just as the intrinsic nature of a taxon explains both the truth of generalizations about its members and why being in the taxon is explanatory, so too the intrinsic nature of an organism explains both the truth of claims about its phenotypic properties and why being that organism is explanatory. The Kripkean intuition that there is an intrinsic component to *Eo*, *Partly Intrinsic Individual Essentialism*, is justified.

Objection: "It is of course the case that the truth of claims about *O*'s phenotypic properties must be explained by an intrinsic, probably largely genetic, property, but why does that property have to be an *essential* property of *O*?"[11]

Reply: *O* has those phenotypic properties *because* it is Benji; it is because *O* is Benji that it has that particular face marking; if it were not Benji it likely would not have had that marking. So the property that *makes* something Benji, whether we call that property an "essence", "nature", or whatever, must *cause* Benji, in his environment, to have stripes; the essence of Benji must explain the place that Benji has in the causal nexus *just because he is Benji*. That's the Sober demand. And the essence of Benji is whatever property, as a matter of actual fact, plays that causal role in an environment.

Some clarification of the intrinsic component of *Eo* is called for. Teresa Robertson and Philip Atkins express the plausible intuition of "modal tolerance of origin": "even if an object could not have had a *completely* different origin from the one it actually had, it could have had *a slightly* different origin" (2016). They note Kripke's claim that the Queen could not have originated from "totally" different gametes (1980: 113). What difference, short of total difference, might Kripke tolerate? The following remark provides insight: "I might have been deformed if the fertilized egg from which I originated had been damaged in certain ways, even though I presumably did not yet exist at that time" (p. 115, n. 57). Kripke is *not* allowing, what would be contrary to *Historical Individual Essentialism*, that he might have come from a *quantitatively* different zygote: the damaged zygote would be the very same zygote that he did come from. Rather, that zygote might have been *qualitatively* different because of damage: it might have had some genetic properties that differ from its actual ones. One wonders, of course, how much difference can be tolerated. Intuitively, if *O*'s zygote had been different in relatively minor ways—say, the nullification of the genetic structure that makes a certain disease likely—the result would still be *O*, but if it were changed massively, it would not be; and the boundary between what is minor and massive is indeterminate. And it may well be the case that this sort of tolerance fits the demands of structural explanations.

So the intrinsic component of *Eo* may tolerate some variation in the zygote at the genetic level. And, we should add, that component would tolerate any amount

[11] The objection is analogous to one, made by Peter Godfrey-Smith, against my argument for *Partly Intrinsic Taxon Essentialism* (2008: 354; 1.3 above). And my present reply is analogous to one I have made to that earlier objection (2021a: 13–14; 2.5 above).

of variation at the molecular level that underlies the genetic level. I think that would be Kripke's view. In any case, it is mine.

4.4 An Argument for *Partly Historical Individual Essentialism*

What about the historical component, *Partly Historical Individual*? The need to support Kripkean intuitions about this is particularly pressing given that some influential philosophers, Graeme Forbes (1986) and Teresa Robertson (1998), do not share them all. On my Kripkean view, three relations are essential to the Queen: (a) she must come from a certain zygote; (b) that zygote must come from certain gametes; (c) those gametes must come from certain parents. Forbes and Robertson do not resist (a) but their modal intuitions count against (b) and (c). Forbes argues as follows:

> Suppose *z* is a human zygote that is formed by fusion of a sperm *s* with an egg *e*. Then one can conceive that scientists synthesize a zygote by building it nucleotide by nucleotide, and happen to use exactly the actual matter of *z* in exactly its actual *z*-configuration. In such a world, *s* and *e* do not exist, or so we can consistently postulate, but it is hard to deny that *z* exists... So *z* exists but does not originate from *s* and *e*, since they do not exist. (1986: 7)

Robertson agrees and claims that others who write on the topic do too (1998: 732, n. 5). Clearly, I do not agree:[12] it seems to me that the synthesized zygote is not *z* precisely because it lacks the right history.[13] But we need more than intuitions.

Once again I offer a two-part argument that is analogous to an earlier one about *Et*, but this time the earlier one is from "Historical Biological Essentialism" (2018a; Chapter 3 above) and argues for an historical component of *Et*. The first part of that argument is that historical explanations of generalizations about the phenotypic properties of members of a taxon, explanations of what led to the taxon having mechanisms that make the generalizations true, advert to the historic component of the taxon's essence. The analogous argument is that historical explanations of the phenotypic properties of an individual organism *O*, explanations of what *led to* there being *O* with the mechanisms that cause those properties, advert to the historical component of *O*'s essence. The historical essence of *O* explains how in its environment it *came to have* its phenotypic properties: it evolved that way because of its history, because of what it inherited.

[12] Nor does Joseph Sartorelli. In a recent article he rightly emphasizes the "importance of biological process in the constitution of continued identity through change" (2016: 1615).

[13] This avoids what Robertson and Atkins (2016) nicely call, "The Recycling Problem": if all the matter that constitutes *z* were recycled into something that was qualitatively identical to *z* it would still not be *z* because it would lack the right history.

In the second, related, part of the earlier argument, I argued that being a member of a certain taxon is explanatory. Why is it? Because the essential nature of a taxon, to be discovered by biologists, causes its members, in their environment, to have their phenotypic properties. What nature? I argued that if our concern is historical, so it is with a nature that led to there being that taxon with those properties, the nature must be historical. We now offer the analogous argument about why being a particular individual organism, say Benji, is explanatory. The essential nature of Benji, to be discovered by biologists, causes him, in his environment, to have his phenotypic properties. And when our concern is historical, with what led to there being Benji with those properties, the nature must be historical. For an organism to be Benji is for it to have the historical essence of Benji. Some aspect of the history of Benji, constituting that historical component of his essence, in combination with the environment, brought it about that Benji exists with his phenotypic features. The same history that (partly) makes the organism Benji causes Benji to be striped. That's why being Benji is explanatory. We have met the Sober demand.

In sum, the demands of historical and structural explanations have yielded two doctrines, *Partly Historical Individual Essentialism*, according to which *Eo* has an historical component, and *Partly Intrinsic Individual Essentialism*, according to which *Eo* has an intrinsic component. On my Kripkean view, the historical component is *O*'s origin in a certain zygote, the result of a certain sperm from a certain male fertilizing a certain egg from a certain female, and so on back through history. And the intrinsic component is the underlying, probably largely genetic, properties that are responsible, along with the environment, for *O* developing its phenotypic properties.

What about the Lucky-Jim worry of section 4.1? The topic of an organism's individual essence should not have been neglected by philosophers of science because that essence plays a causal role in structural and historical explanations of the organism's phenotypic properties. One need not call this property of an organism an "essence", of course, but whatever one calls it the property, along with the environment, explains the phenotypic properties. So it is interesting.

One further matter needs to be considered briefly.

4.5 Haecceitism

Suppose that *O* has an identical twin *O**: *O** developed from the same zygote as *O* but the zygote split yielding two embryos and hence two organisms. So *O** shares with *O*, at least, the *qualitative* intrinsic component of *Eo* and the historical component up to the splitting. Yet clearly *O* and *O** differ and so must differ in their individual essences. What does that difference consist in? At least, they differ in their histories after the splitting. But is that *all* the difference?

Clearly the zygote *might not have split* in which case there would have been just one person and not both *O* and *O**. Could that person be *O*? Could it be *O**? If I understand the traditional metaphysical doctrine of *haecceitism* (which I may not), that doctrine gives positive answers. For, according to that doctrine, worlds can differ *non*-qualitatively without differing qualitatively. So in one world the person that results from the non-splitting zygote might have the intrinsic non-qualitative property of *being O* and nothing in that world has the intrinsic non-qualitative property of *being O**; whereas in another world, vice versa. I have nothing of interest to say on this issue.[14] However, so far as I can see, whatever one says can be accommodated by *Intrinsic Individual Essentialism*.

I have said what I take *Partly Historical Individual Essentialism* and *Partly Intrinsic Individual Essentialism* to be committed to. These two theses, understood in this Kripkean way, are my account of *Eo*. So far as I know, the biology literature does not present a rival account.

We are ready now to turn to the issue of *Essential Membership*.

4.6 *Essential Membership*

Intrinsic components: Now if the controversial *Partly Intrinsic Taxon Essentialism* is false then the intrinsic component of *Eo* is irrelevant to *Essential Membership*. For, if no intrinsic property of *O* is essential to *O*'s being a member of taxon *T*, then *O*'s having an intrinsic property essentially must be irrelevant to its essential membership. But if *Partly Intrinsic Taxon* is true, then *Essential Membership* requires that *O*'s having the intrinsic component of *Eo* entails *O*'s having the intrinsic component part of *Et*. If *Et* and *Eo* have the intrinsic components described above, then this entailment holds.

The entailment holds because the intrinsic component of *Eo* features in the structural explanation of the phenotypic properties of *O* (4.3), and that of *Et* features in the structural explanation of phenotypic properties that are *common* to members of *T*. Thus, among all the properties of *O* caused by the intrinsic underlying component of its essence *Eo*, together with its environment, are the set of properties that are, according to the generalizations about the phenotypic properties of members of *T*, common to those members in the environment they share with *O*. But the underlying properties of members of *T* that, together with that shared environment, cause those members to have that set of common properties constitute the intrinsic component of *Et*. So *O*'s having *Eo* entails *O*'s

[14] However: (i) I prefer negative answers to the questions: the person could not be determinately *O* or determinately not *O**; rather, the person would be *either O or O** but there would be no determinate matter of fact which; (ii) I clearly reject "extreme *haecceitism*" according to which *haecceities* are the *only* essential properties of individuals.

having that intrinsic component of *Et*. Consider the tiger Benji, for example. Set aside Benji's peculiarities and consider his property of being striped, a property typical of tigers in his environment. A part of the intrinsic underlying component of the individual essence of Benji explains why *he* has the property of being striped. That part of the intrinsic underlying component of the essence of tigers also explains why *tigers* have that property of being striped.[15] The "sum" of all such parts of Benji's essence, is the intrinsic component of the essence of tigers.

So if the controversial *Partly Intrinsic Taxon* is true, then *Essential Membership* will be true provided that *O*'s having the historical component of *Eo* entails *O*'s having the historical component of *Et*. And, of course, if that controversial doctrine is not true, the fate of *Essential Membership* depends entirely on this proviso about historical components. So, let us turn to the historical components.

Historical components: Whether or not an organism must have the historical component of all of its taxa, particularly of its species, obviously depends on what those components are. There is a problem about this, as noted in section 4.1, because biologists and philosophers of biology have had strikingly little to say about what precisely these historical essences are, and what they have said is not plausible. However, I have urged that the essential history is of organisms of *a certain intrinsic kind* evolving into organisms of *a certain other intrinsic kind*, until we reach the taxon in question (2018a: 6; 3.5 above). This hypothesis clearly presupposes the controversial *Partly Intrinsic Taxon Essentialism* and so it is sure to be controversial too.

Nonetheless, let us go with this hypothesis about the historical component of *Et* for a moment. On my Kripkean view, the historical component of *Eo* is *O*'s origin in a certain zygote, the result of a certain sperm from a certain male fertilizing a certain egg from a certain female. And this component includes *O*'s parents and constitutive zygotes, sperm, and eggs, and so on back through *O*'s family tree (4.4). Now if *O* is a member of *T*, this historical component of *Eo* will, on my account, exemplify the historical component of *Et*. For, that component of *Et* consists in *T* having the intrinsic component, *P*, of *Et as a result of* having evolved from another taxon, *T**, with a certain different intrinsic component, *Q*. We have just seen that *O* has the intrinsic component of *Et* and hence *P*. And *O* has *P* because of a history that includes ancestors that had *Q*. Similarly, for the part of the historical component of *T* that includes the evolution of *T** from *T***, and so on back to the beginning of the tree of life.

Given our earlier discussion of the intrinsic components, this would establish *Essential Membership* but for the following concern. This argument for *Essential Membership* depends on my hypothesis about the historical component of *Et*,

[15] This is a bit too simple because of the often disjunctive nature of the underlying cause of a phenotypic property (Devitt 2021a: 12–13; 2.4.6 above).

likely to be controversial. But then it remains to be seen whether a plausible alternative to this view can be produced. For an alternative to be plausible, note, it has to carry the explanatory burden; the Sober demand. Meanwhile, the best I can do is to *predict* that on any such alternative, it will be a consequence of our account of the historical component of *Eo* that, if *O* is a member of *T*, *O* must have that historical component of *Et*: *Essential Membership* will hold. I am confident about this prediction because of the very richness of the historical component of *Eo*: so much of the history of *T* is to be found in the individual essence of each of its members.

Finally, I turn to LaPorte's objections to *Essential Membership*. I have noted that the higher taxa tend to get ignored in discussions of essentialism. LaPorte's discussion (1997) is an exception. He offers objections to *Essential Membership* in species *and* in higher taxa. I shall consider the ones for species first.

4.7 Objection 1: The Interbreeding and Ecological Approaches to Species

LaPorte starts his objections as follows:

> I argue... that organisms do not essentially belong to the *species* to which they belong. To show this, I do not want to assume any particular theory about what determines the boundaries of species... biologists do not agree on what species are, about what it is that makes the members of a species members of the species. (1997: 101)

This looks like a promising start: to assess whether an organism is essentially a member of its species—*Essential Membership*—we clearly need to know what it is for the organism to be a member: we need an answer to Mayr's *taxon* problem. But LaPorte then turns immediately to the species concepts for answers. As he notes, these concepts "tend to fall into three camps: the interbreeding approach, the ecological approach, and the cladistic approach". He claims that "each... sacrifices the essentiality of species membership" (1997: 101). Yet the species concepts provide answers to Mayr's *category* problem and, as I have been emphasizing (1.5–1.7, 1.9, 2.3, 2.6), throw little light on the taxon problem. LaPorte's discussion is a nice illustration of the mistaken role that species concepts have played in discussions of biological essentialism.

I shall consider LaPorte's objection from the perspective of the interbreeding and ecological approaches in this section, that from the perspective of the cladistics approach in section 4.8. In section 4.9, I address an objection based on Kitcher's ingenious case of "dumbbell allopatry". Finally, in section 4.10, I consider LaPorte's objection to *Essential Membership* in higher taxa.

Laporte considers "a large population of organisms, from which a small population splinters off". This small population "takes up a new ecological niche" and becomes "reproductively isolated" from the larger population (1997: 101).

> Both of the above approaches [ecological and interbreeding] would consider the two branches to be distinct species. Organisms of the smaller branch do not belong to species "*A*", whose members constitute the larger branch Yet this could be a plainly contingent matter. Had the members of the little branch not taken on a new niche, or had there not been reproductive isolation... between them and the members of *A*, they would belong to species *A*, given the species concepts in question. And surely it is possible that members of the side branch should have remained the original niche. (p. 102)

This is confused. (1) It is indeed possible that *the organisms that splintered initially from A* should have remained in *A*. Then there would not have been the new species, "*B*". But those organisms that splintered were *A*s not *B*s. There *were* no *B*s until speciation had occurred, many generations later, as a *result* of the initial splintering. Speciation does not occur overnight. (2) It *is not* possible that *the members of B* should have remained in *A*. They never were in *A*: they were the result of generations of breeding in a new niche isolated from *A*. *B*s owe their very existence to the initial splintering. It is true that the splintering that led to speciation might not have happened and then there would have been no *B*s, just *A*s. But it is not true that, once speciation occurs and there were *B*s, that any of them might have been *A*s.

4.8 Objection 2: The Cladistic Approach to Species

LaPorte turns next to the cladistic approach. This yields an objection to Membership that is of quite general interest and so I will consider it at some length. The objection arises from a common feature of the popular cladistic concept ("CC"). According to CC, a species "is a lineage of organisms between two speciation events, or between speciation event and one extinction event" (LaPorte 1997: 102). The common feature that causes the trouble was proposed by Willi Hennig (1966) and is described by LaPorte as follows:

> a species goes extinct whenever it sends forth a new side species. This is so even if the lineage undergoes no change after sending the side branch, so that earlier members are indistinguishable from later ones. (1997: 103)

In brief, a species *cannot* survive a split. Suppose then that *O* is actually a member of *A* and *A* is a species on this version of CC. Then *had* a daughter *C* split off from

A before *O*, but without any effect on *O*'s ancestors, *O* would not have been a member of *A* but rather of a successor species *B*. That is, even though this split would have had no effect at all on *O*'s history or intrinsic properties, it would have had the consequence that *O* would not have been in *A* but rather in *B*. So *O* is not essentially a member of *A* and *Essential Membership* for taxa that are species is false.

Now we should accept immediately that *if* any taxon that concerns us here is a species *according to a CC that includes Hennig's troublesome feature*, then *Essential Membership* does not hold for it.[16] So, to that extent, but only to that extent, *Essential Membership* would have to be qualified. But *are* any of our taxa of that sort? This is not easy to answer. Attempting to answer it is of considerable independent interest because of the influence of Hennig's view.[17]

(1) We need to start our answer by considering the nature and status of CC. As a species concept, CC is a theory of what it is for a taxon to be a species (rather than, say, a subspecies or genus). Now as we noted in section 1.4, and LaPorte himself has just noted, CC has *lots* of company. As Kitcher remarks, the species problem is "one of the thorniest issues in theoretical biology" (2003: xii). Perhaps no species concept is objectively right: "There is no available fully objective ranking criterion for species" (Baum 2009: 76). One notable response to this controversy has been to reject the view that there is *just one* good species concept, to reject "species monism", and to urge "species pluralism", the view that there are *many* (Kitcher 1984; Ereshefsky 1998; Dupré 1999). In brief, it is uncontroversial that species concepts are controversial.

One bit of the controversy is particularly pertinent because it concerns the very feature of CC that threatens *Essential Membership*, the feature that a species cannot survive having a daughter. I have noted that CC is frequently understood as having this feature, but it is certainly not always so understood. Edward Wiley (1981) rejected this feature and it is "contentious ... in evolutionary theory" (Pedroso 2012: 186). Peter Godfrey-Smith points out that the feature "can be avoided by saying that if one of the new branches is much larger and the other is a small 'budding,' something that probably often happens in nature, then the old species has lived on in the larger branch" (2014: 105).

[16] But note that this view of species does not count against *Essential Membership* for genera. Thus, suppose that species *A* is a member of genus *T*. Then, had *A* split before *O*, *O* would not have been a member of *A* but it still would have been a member of *T*.

[17] Hennig's view is that speciation is the splitting of an old species into two new species. So, such a split is *sufficient* for a species to end; that is our troublesome feature. But such a split is also *necessary* for a species (that has descendants) to end: there can be no anagenesis; in that respect, a species is "open-ended". This has been taken to be at odds with *Partly Intrinsic Taxon Essentialism* (Sterelny and Griffiths 1999: 7). Initially I agreed that it was (2008: 369) but changed my mind later (2010: 238, n. 40; 1.9 n. 43 above). In any case, we should ask the question: "Are any actual taxa we take to be species open-ended in this way?" An argument like the one to follow suggests not.

It helps to understand the controversy over species concepts to consider what motivates the species pluralists. According to Kitcher, many concepts "can be motivated by their utility for pursuing a particular type of biological inquiry" (1984: 118). Kyle Stanford puts the point thus: "certain explanatory demands are *inextricably bound* to certain species concepts" (1995: 72). And there are many different, but equally legitimate, types of biological inquiry and explanatory demands: "we have independent and legitimate explanatory interests in biology which require distinct concepts of species" (p. 76). The key message that we should take from the pluralists is that a species concept is motivated by the explanatory role of being a species according to the concept: a taxon should be in the **Species** category specified by the concept because it plays a certain causal role *in virtue of* being in that category. And the controversy arises because there is uncertainty and disagreement over what explanatory role, or roles, is supposed to be played by being in a **Species** category. Indeed, perhaps it plays none.

(2) A species concept, including CC, is clearly meant to be normative, saying how biologists *ought to* classify taxa as species, given the explanatory role of being in a **Species** category. But I take it that a concept is also supposed to be descriptive, saying how biologists *do* classify taxa as species, for the most part at least. And only if CC, *with the troublesome feature*, is descriptive could it bear immediately on the doctrine *Essential Membership*, for that doctrine is about *actual* taxa. Do we have any good reason to suppose that that troublesome and controversial feature *is* descriptive of biological practice? Do biologists actually have that feature in mind when they classify a taxon as a species? Kim Sterelny thinks that they often do not: "Some, perhaps most, evolutionary biologists take speciation to occur only when there have been intrinsic changes" (1999: 130). So for many evolutionary biologists having a daughter is *not sufficient* to end a species. And what about *non*-evolutionary biologists concerned with structural rather than historical explanations?

> Classification in biology has a life of its own. Biologists in areas only tangentially connected to evolutionary theory, such as ecologists, ethnobotanists, or ethologists, need to classify organisms, as do foresters, conservationists, gatekeepers, and herbalists... for many, perhaps even most groups of organisms, evolutionary considerations are of little or no use for classificatory purposes.
>
> (Dupré 2002: 82)

Given the controversy described in (1), there has to be doubt whether *any* particular one of the given species concept is *normatively* true or even, should pluralism be correct, part of the truth. Even if a concept is normatively true, or part of the truth, we have just seen that there has to be further doubt whether it is *descriptively* so and hence potentially relevant to *Essential Membership*. But suppose a concept *is* descriptively true or part of the truth, *how* might it bear on

Essential Membership? This is difficult. In thinking about it, it is crucial to keep the taxon/category distinction in mind (1.5, 2.3). A species concept offers a solution to the *category* problem by telling us what is essential *for a taxon T to be a species*. Yet what is relevant to *Essential Membership* is a solution to the *taxon* problem, the problem of saying what is essential *for an organism to be a member of T*. So how can a species concept bear on *Essential Membership* even if it is true?

(3) Suppose that biologists *take T* to be a species. So *T* is a taxon covered by *Essential Membership*. Suppose next that, according to some given species concept *SC*, *T is in fact* a species. How could its being so bear on what is essential for an organism to be a member of *T*? Only if *SC* played a determining causal role in the way biological taxonomists classify organisms as members of *T*. *SC* has to be descriptive *in that respect*. How could it be? There seem to be two requirements on its being so: (i) taxonomists must embrace that very *SC* in classifying *T* as a species; (ii) this embrace must partly determine what counts as being a member of *T*. These are demanding requirements.

First, one wonders whether the typical taxonomist embraces any particular species concept. Given the level of controversy over these concepts, one suspects that many taxonomists suspend judgment on this vexed matter. Second, taxonomists who do embrace a concept may embrace one of the many ones other than *SC*; or they may vary among themselves in the concepts they embrace. The earlier cited (2.3) survey of the opinions of 193 biologists in the EU and USA is relevant to these two points. The survey claims that these opinions provide

> strong evidence against monism, since they clearly indicate that there is more than one species concept in use in contemporary biology.... A further blow to monism is... [that] 80% of our participants think it false that there is one species concept applicable to all fields in biology... [and] more than a half of our participants think that monism is not even a desirable position in biology.
>
> (Pušić et al. 2017: 196)

Third, even if taxonomists did all have *SC* in mind as they identify and name *T*, it is far from clear what if any role that would play in determining the extension of *T*. Indeed, it is far from clear what mental states *in general* play a role in determining extensions. It seems to me quite likely that the extension of *T* is determined independently of any theorizing that biologists may engage in about the nature of species.[18] Rather than playing a role in determining the extension of *T*,

[18] Contrast this with what Haber has to say about how a researcher uses a type specimen to fix the extension of a taxon: "codes of nomenclature are silent on the criteria of belonging to a taxon, i.e., it is left to individual researchers to decide what species concept to apply to determine the taxonomic boundaries about a type specimen" (2012: 769, n. 3).

these concepts may simply be theories of *T*, the extension of which has already been determined.

We are confronting the deep and difficult matter of reference fixing. We must take note here of the revolution in the theory of reference started by Kripke (1980), particularly the devastating "ignorance and error" arguments against description theories. So we should be reluctant to suppose that, when people introduce '*T*' to name a taxon, its reference is determined by some description that they associate with the term. Rather we should look for the reference of '*T*' being determined by some sort of causal "grounding" in specimens of *T*. Something that goes on in those groundings, including mental states of the grounders, determines how the reference of '*T*' is projected from that specimen to all other *T*. But *what* does this job? Any specimen is a member of *many* kinds. In virtue of what is the reference projected from the specimen *qua* member of one of those kinds rather than *qua* member of another? At this point we don't know how to solve this "qua-problem"; see Devitt and Sterelny (1999: 88–101) and Devitt and Porter (2021).

(4) So, even if a certain species concept was descriptive of the way biologists classify taxa as species, we have reason to doubt that this concept would be relevant to the issue of what is essential for an organism to be a member of a given taxon thought to be a species. Consideration of taxon reclassification adds to this doubt.

It is common in biology for some taxon *T* thought to be in one category to be reclassified in another. We gave the examples earlier (2.4.1) of the Sumatran tiger and the African forest elephant that were thought to be subspecies but were reclassified as species. But let us take an example of a taxon, the British red grouse, that was thought to be a distinct species of the genus *Lagopus* and was reclassified as a subspecies: it is one of nineteen subspecies of *Lagopus lagopus* that are scattered over many countries. I am here following the standard way of describing this sort of reclassification in biology. Thus, *Wikipedia* describes the change as follows: "The distinctive British subspecies *L. l. scoticus* (red grouse) was once considered a separate true British species but is now classified as a sub-species". The clear "identity assumption" here is that *one and the same taxon*, the British red grouse, used to be classified as a species and is now classified as a subspecies. How could this identity assumption be reconciled with the view that the extension of the taxon is partly determined by whether or not the taxon is a species according to a certain species concept?

The original classification as a species, and the reclassification as a subspecies, must each have arisen, of course, from taxonomists applying a species concept, even if only an implicit one. Now the species concepts applied in classification and reclassification are either the same or different. Consider each possibility in turn.

First, suppose that the species concepts applied are the same. So, if the concept really partly determined the extension of the taxon, then the taxonomists who reclassify the taxon should think that the taxon lacks any members. So, contrary to what we have thought for decades, there are not, and never were, any British red grouse! For there to be any such grouse, the taxon would have to fit the concept's description of a species and the taxonomists have just concluded that this group of birds does not. The taxonomists obviously still think that all those birds *exist* but, contrary to the identity assumption, they should not think that any of those birds *are British red grouse*. I take it as rather obvious that they do still think of those birds as British red grouse, despite the reclassification.

So if there is hope for the reconciliation of the identity assumption with the alleged extension-determining role of the species concepts it must be found in situations where the concepts applied in classification and reclassification are different. Indeed, reconciliation *is* possible in such a situation. For, taxonomists in the reclassification can accept that the taxon still fits the original concept's description, so can accept that there really are British red grouse. What they now think, as a result of the concept change, is that that very group of grouse do not constitute a species.

We should surely assume that in taxon reclassification taxonomists are not forced to conclude that a group of organisms thought to form a certain taxon do not in fact do so. Rather they conclude, simply, that that very taxon of actual organisms is not, as previously thought, say, a species. If this is so, and species concepts really do partly determine a taxon's extension, then our discussion shows that reclassification from species to subspecies is *possible* only where taxonomists change their species concept. That is very implausible. Indeed, it seems plausible that taxonomists often reclassify without any change in their species concepts simply as a result of more information about taxa. For example, it is plausible to suppose that the reclassification of the British red grouse came about simply from studying it and the other eighteen subspecies. *So, we have evidence here that whether or not the taxonomists who identified and named the taxon thought of it as a species according to a certain species concept has no effect on its nature and hence no bearing on what is essential to being a member of a taxon.*

It is time to sum up the discussion in this and the last section. I had earlier argued for *Essential Membership* in species (4.6). Laporte rejects this doctrine, claiming that "it is doubtful that *any* account according to which species are historical entities (lineages), results in organisms essentially belonging to their respective species" (1997: 104). In section 4.7, I considered LaPorte's objection to *Essential Membership* based on interbreeding and ecological species concepts. He considers a situation where a new species *B* splinters from an old species *A*. I argued that his objection confuses the truth that *A* might not have splintered and so there would have been no members of *B*, with the falsehood that the actual members of *B* might have been members of *A*.

In the present section, I have looked critically at Laporte's objection to *Essential Membership* based on a version of CC with the troublesome feature that a species cannot survive having a daughter. I started by conceding that *if* any taxon that concerns us here is a species *according to a CC that includes this feature*, then *Essential Membership* does not hold *for it*. And it has to be allowed that there *may* be such taxa among those commonly thought to be species (but not among others). However, I have adduced four reasons for thinking that it is unlikely that there are such taxa. First, any species concept is controversial and the troublesome feature of CC is particularly controversial. Second, even if CC with that feature is true, or part of the truth, as a normative doctrine, it could bear on *Essential Membership* only if it describes how biologists actually classify taxa as species. Third, even if it is descriptive, what it describes may not play any role in determining the membership of taxa thought of as species. Fourth, the common practice of reclassifying a species as a subspecies, and vice versa, is at odds with the view that there are taxa that have their membership determined by any species concept.

4.9 Objection 3: Kitcher's "Dumbbell Allopatry"

Kitcher presents a wonderfully ingenious problem for "all versions of the thesis that species are historical entities... It arises from the simple possibility of 'dumbbell allopatry' as a mode of speciation" (Kitcher 2003: 152). This possibility also raises a problem for *Essential Membership* in species, as LaPorte points out. LaPorte describes Kitcher's possibility as follows:

> Kitcher imagines... an evolving population, which divides at *t* into equal halves. By *t′* the branches have diverged sufficiently to constitute separate species, whatever the criterion used to determine that: reproductive isolation, or something else. The divergence stops at t′. At no time is the distance between the ancestral lineage and either branch sufficient for speciation; speciation occurs only because of the distance attained between the two branches. (1997: 104)

The basic problem for *Essential Membership* arises from this counterfactual:

> A cataclysm could have made things otherwise: for had either branch been prevented from evolving until t′, there would not have been sufficient distance between the branches for speciation to have occurred. (p. 105)

Here is my best attempt to state the alleged problem posed by this counterfactual. Let us call the species before *t*, "*X*", and the two species after *t′*, "*Y*" and "*Z*". The problem for *Essential Membership* is that had *Z* not become a distinct species—perhaps the organisms that formed the founder population for *Z* were wiped out before speciation—then there would have been no new species *Y*: the divergence of the organisms that actually constitute *Y* from those that constitute *X* is not *alone* enough to make them a new species in that counterfactual situation.

So, *O*, which is actually a member of species *Y*, would not be so in that counterfactual situation: it would be a member of species *X*. So, *O* is not essentially a member of its species *Y*.

My response hinges on distinguishing the question whether *O* is essentially a member of a certain group and the question whether that group is a species. Once again, attention to the taxon/category distinction is crucial. Applying my earlier conclusions, I consider first the relevant features of the actual situation described by Kitcher, and then the counterfactual situation:

Actual situation:

1. *O* is essentially a member of group Y.
2. The members of *Y* form a species.
3. So, *O* is essential a member of its species.

Counterfactual:

4. *O* is essentially a member of the group that is the combination of *X* and *Y*: *X&Y*.
5. The members of *X&Y* form a species, but the members of *Y* alone do not.
6. So, *O* is essential a member of its species.

So Essential Membership holds in both the actual and counterfactual situation. So where did my best attempt at stating the alleged problem for *Essential Membership* go wrong? It is indeed true that, in the counterfactual situation, there would have been *no new species Y*. Nonetheless, there still would have been group *Y* with *O* as a member. The crucial change in the counterfactual situation is that *Y* *would not then be a species*. Furthermore, *O* would be a member of a species formed by the members of *X&Y not simply of X*, as the best attempt claimed.

Consider the steps. The argument for *Essential Membership* (4.6) applies to taxa of any category (and much else besides). So that argument establishes steps 1 and 4. The problematic steps are 2 and 5, for they reflect what concerns Kitcher. For, according to 2 and 5, whether or not the members of *Y* form a distinct species depends on what is going elsewhere: on whether or not the species *Z* comes into existence. This flies in the face of a "principle" that Kitcher thinks we should accept:

> (*) A proposal to count lineage-stages as stages of the same species should depend only on the intrinsic properties of and direct relations among those stages. It should give the same results in cases which differ only in the existence or properties of organisms occupying a different branch of the lineage. (2003: 150)

So dumbbell allopatry does not pose a problem for the essential membership of *O* in a certain group but for the view that that group is a species. That is a problem for a species concept. Perhaps 2 and/or 5 take a wrong stand on this. So be it: species concepts are not my business. *O* is an essential member of *Y* in the actual situation and of *X&Y* in the counterfactual situation, whether or not they are species.

4.10 Objection 4: Higher Taxa

We turn finally to the higher taxa. LaPorte asks, "Could individual mammals be such that they are mammals in all possible worlds?", and responds that "here, too, essentialism is hard pressed for a toehold" (1997: 106). I disagree, of course. My two-part arguments that taxa (Chapters 1–3) and individuals (4.3–4.4) have intrinsic and historical components to their essences yield an argument for *Essential Membership* that is quite general, applying to *O*'s membership in all its taxa not just its species (4.6).

Indeed, suppose that *Essential Membership* in species is right. So the Queen is essentially a member of *Homo sapiens*. I remarked earlier:

> Where taxon *T* is in a lower category than taxon *T** and falls within *T**, the essence of *T* must include the essence of *T** but not vice versa. The underlying essential property of *T** that explains why the members of *T** have phenotypic property *P* explains why the members of *T* have *P*; hence that underlying property is part of *T*'s essence. (2.3)

The Queen is essentially a *Homo sapiens* and so has the essence of that species. That essence includes the essence of *Homo* (genus). So she is essentially a *Homo*. The essence of *Homo* includes the essence of hominids (family). So she is essentially a hominid. And so on: she is essentially a primate (order); and, contrary to LaPorte, she is essentially a mammal (class). We can go further: she is essentially a chordate (phylum) and essentially an animal (kingdom).

What does LaPorte say against *Essential Membership* in higher taxa?

> There are three main competing modern schools of classification: phenetic taxonomy, evolutionary taxonomy, and cladism.... my aim is to show that none of the schools is consonant with [*Essential Membership*], and hence to show that [*Essential Membership*] is out of touch with biological systematics.
> (1997: 106)

What exactly are these schools of taxonomy or, as it is often now called, of "systematics", schools *of*? There are various definitions in the literature. Here are two of taxonomy:

> A field of science (and major component of systematics) that encompasses description, identification, nomenclature, and classification. (Simpson 2010: 575)

> The science of classification, in biology the arrangement of organisms into a classification. (Kirk et al. 2008)

It is important to distinguish two sorts of classification and naming that concern taxonomy, one a taxon issue, the other, a category issue:

> *Taxon*: The identification and naming of a group of organisms as a biological taxon. In virtue of what should that group be classified as a biological taxon?
>
> *Category*: The identification of the category of a biological taxon. In virtue of what should that taxon be classified with its Linnaean rank?

And, in light of what has been emphasized before (1.5–1.7, 1.9, 2.3, 2.6, 4.7), we can say immediately that, insofar as the three schools of classification are talking about the *Category* issue, they will throw little light on our essentialism issues. So, our concern is only with what the schools say on the *Taxon* issue.

I start with the cladistic school. In discussing *Essential Membership* in species, we saw that the cladistic species concept did yield one, sort of, exception (4.8). We allowed that *if* organism *O* is in a taxon *T* that is a species according to a cladistic concept that includes the troublesome Hennig feature, *then O* is not essentially a member of *T*. The troublesome feature was, of course, that "a species goes extinct whenever it sends forth a new side species" (LaPorte 1997: 103). Turning now to higher taxa, if *O* is not thus essentially a member of *T*, which is a species, then it is not essentially a member of any of the higher taxa that include *T*. That, in a nutshell, is LaPorte's cladistics objection to *Essential Membership* in higher taxa:

> For cladism, a "stem" species, which gives rise to different successor species that together comprise a single taxon, is itself part of the taxon (Hennig 1966, 71–72). For example, some particular species of *Archaeopteryx* gave rise to the many existing birds. Hennig (pp. 71–72) counts that species as the stem species of the birds and hence counts it itself as a species of bird.
>
> But if cladism is right, it is a contingent matter that any individual member of the relevant *Archaeopteryx* species is a bird.... For consider: it could have happened that the *Archaeopteryx* also budded forth a non-bird taxon at any time before the first branch was sprouted.... In that case, those organisms that exist prior to bird offshoot... and that, as things actually are, comprise the stem for the Aves, would fail to count as birds.... So a contingent event's occurrence or non-occurrence determines membership in the Aves: whether the side branch does or does not bud off decides the boundaries of the taxon. (1997: 109–10)

This objection to *Essential Membership* in higher taxa depends on there being taxa that are species according to the cladistic concept with the troublesome

Hennig feature. So, I can be quick in responding to the objection. I have given four reasons for thinking that it is unlikely that there are such taxa (4.8). In brief, first, this species concept is controversial. Second even if the concept is normatively true, it needs to be descriptively true to bear on *Essential Membership*. Third, even if it is descriptive, what it describes may not play any role in determining taxon membership. Fourth, reclassification practices are at odds with the view that there are taxa that have their membership determined by any species concept.

Turn now to LaPorte's phenetic objection to *Essential Membership* in higher taxa:

> This method of taxonomy... takes into account the various characteristics observed to belong to organisms, and gives these characteristics equal weight. Then, it makes use of computers to provide groupings by overall similarity.
>
> But if our taxa are defined phenetically, essentialism about kind membership loses. (1997: 106–7)

This is surely so. For, according to the phenetic definition, the essence of a kind is, in effect, a cluster of observed characteristics. Yet, as LaPorte says, no one supposes that any of these characteristics are essential to an individual (p. 107).

The objection to this objection is easy and obvious: the phenetic view of taxon essence is quite wrong, for the reasons given in Chapters 1–3. The essence is constituted by the intrinsic underlying and historical properties that are, along with the environment, *causally responsible* for the observed cluster; cryptic species are made up of phenetically similar organisms that are nonetheless in different species because of underlying differences (2.4.1). As already noted (1.4), the appeal of the phenetic approach to its proponents is that it is "operational". In my view (1997), "operationalism" in science, is rightly discredited, mistaking the observable evidence of a scientific reality for that reality itself.

This talk of observable evidence draws attention to issues in taxonomy other than the above two about the natures of biological taxa and categories. Each of these "constitutive" issues generates an "epistemic" issue. *How do we tell* whether a group of organisms is a biological taxon worthy of a name? If it is, *how do we tell* what category it is in? Some of the controversy between taxonomic schools can be seen as over these epistemic issues.

Turn finally to LaPorte's evolutionary objection:

> Evolutionary taxonomy... uses both phenetic and phylogenetic information in classification.... evolutionary taxonomy can, like pheneticism, allow individuals existing simultaneously, or at different times, to fall into separate camps on the basis of plainly accidental characters. A simple example will suffice to illustrate the point. Birds originally evolved from the reptiles. But, of course, evolutionary change is not instant.... The evolutionary taxonomist must... draw a line, on

> the basis of phenetic principles, somewhere between present-day birds, with all of their avian characters, and the ancestors that first displayed one of these characters. So at some point along the line, some newly acquired character (flight, perhaps) will distinguish the first birds from the birds' last ancestors.
>
> But surely there is no guarantee, nor even a likelihood, that the characteristic that separates the one taxon from the other is essential to those who possess it. Suppose it is flight, for example. No essentialist will want to say that the very identity of an individual rests on its ability to fly. (1997: 108–9)

The taxonomist is indeed faced with the problem of drawing a line between a taxon and its immediate ancestor. And given that evolution is not instantaneous we know that there is no determinate matter of fact about where the line is to be drawn; see discussions of "Gradual Change", "Indeterminacy", and "Arbitrariness" in sections 1.10 and 2.4.7. But we should not follow the evolutionary taxonomist in drawing this indeterminate line "on the basis of phenetic principles". The characteristic that distinguishes birds from reptiles is not the observed ability to fly but the underlying genotypic property that partly causes many birds (but not emus or penguins) to fly. That is the property that is essential to the kind **Bird** and to each individual bird. *Essential Membership* survives.

The view on the *Taxon* issue that has run through this book emphasizes *explanation*: a group of organisms should be identified and named as a biological taxon iff those organisms share a nature/essence that is structurally and historically explanatory of their phenotypic properties.[19] This view has elements of all three schools. The cladistic school emphasizes phylogenetic properties. These are the focus of the view's historical explanations. The phenetic school emphasizes phenotypic properties. These are explained by the view's structural and historical explanations. The evolutionary school emphasizes both the phylogenetic and the phenotypic.

I have presented no view on the *Category* issue but will do so (6.7; 2023).

4.11 Conclusion

A few philosophers of biology have recently rejected *Essential Membership*, the doctrine that if an individual organism belongs to a taxon, particularly a species, it does so essentially. But philosophers of biology have not addressed the broader issue, much discussed by metaphysicians, of what *is* essential to the organism. In this chapter, I have addressed this issue from a biological basis.

[19] There is an issue of *how* explanatory (6.5.2).

I first presented (4.2) the Kripkean view that an organism has a partly intrinsic underlying essence, *Partly Intrinsic Individual Essentialism*, and a partly historical one, *Partly Historical Individual Essentialism*. I then argued (4.3–4.4) for these doctrines in turn along lines analogous to those of earlier arguments for similar views about taxa, *Partly Intrinsic Taxon Essentialism* (Chapters 1–2) and *Partly Historical Taxon Essentialism* (Chapter 3). All these arguments appeal to the explanatory concerns of biology.

After a brief discussion of haecceitism (4.5), I turned to *Essential Membership*. My conclusions about the essences of individuals and taxa yielded an argument for *Essential Membership* (4.6).

Finally, I addressed LaPorte's objections, starting with three objections to *Essential Membership* in species. Objection 1 stems from the interbreeding and ecological species concepts. I argued that the objection is confused (4.7). The more interesting objection 2 stems from a cladistics species concept with the troublesome feature that a species cannot survive having a daughter. I give four reasons for doubting this objection (4.8). Objection 3 is based on Kitcher's ingenious case of "dumbbell allopatry". I argued that this case does not pose a problem for *Essential Membership* but only for species concepts (4.9). Objection 4 concerns *Essential Membership* in higher taxa. I argued against LaPorte's claim that the three main schools of taxonomy/systematics count against *Essential Membership*.

I hope to have set to rest the Lucky-Jim worry: the topic of an organism's individual essence is worthy of interest.

5
Type Specimens and Reference

5.1 Introduction

In the last chapter, I argued for *Essential Membership*, the doctrine that if an individual organism belongs to a taxon, particularly a species, it does essentially. This goes against what seems to be the implicit consensus among philosophers of biology. In an ingenious and provocative paper, “Individualism, Type Specimens, and the Scrutability of Species Membership”, Alex Levine sides with the consensus: “species membership, by which I mean the relation that connects a given organism, *o*, with the species *S* of which it is part, is a fundamentally contingent matter” (2001: 333).[1] But then he finds this contingency in conflict with the role of type specimens in biology. He points out that “naming a species requires collecting and preserving one, or at most a very few specimens of the species in question” (p. 327). David Hull has the following view of this practice:

> The sole function of the type specimen is to be the name bearer for its species. No matter in which species the type specimen is placed, its name goes with it.
>
> (Hull 1982: 484)

As we shall see in section 5.2, Levine takes Hull’s view, together with the “rigid designation” theory of reference, to entail that any organism selected as the type specimen for a species is necessarily a member of that species. This generates the conflict that Levine sums up neatly as follows: “*qua organism*, the type specimen belongs to its respective species contingently, while *qua type specimen*, it belongs necessarily”; he finds this “paradoxical” (2001: 334).

What precisely is Levine’s necessity thesis about type specimens? Joseph LaPorte (2003) has clarified this question. He starts with the following statement

[1] Levine rejects *Essential Membership* because of what he takes to follow from the Ghiselin-Hull doctrine that species are individuals:

> If species are individuals, not natural kinds or even “spatiotemporally unrestricted classes”, then the membership of a given organism in a given species is always only a contingent matter of fact, never a necessity. This is not an unforeseen consequence of individualism, but rather a fundamental feature of the doctrine as conceived by its authors. (2001: 330)

In my view, individualism is largely irrelevant to essentialism issues (2008: 348; 1.1 above).

Biological Essentialism. Michael Devitt, Oxford University Press. © Michael Devitt 2023.
DOI: 10.1093/oso/9780198840282.003.0005

of the thesis: 'It is necessary that any species with a type specimen contains its type specimen'. He points out that such statements have two readings:

> The *de dicto* reading of the statement in question would typically be expressed thus: "Necessarily, any species with a type specimen contains its type specimen." The *de re* reading would be expressed: "Any species with a type specimen necessarily contains its type specimen." (p. 586)

Laporte thinks that although the *de dicto* reading is true (p. 587), the *de re* one is not, and this resolves the paradox. The first major concern of this chapter is to argue that the *de dicto* reading, which I shall call "*Levine's Thesis*", is false. That is my conclusion *C1*, argued for in section 5.3.

LaPorte's response to Levine's alleged paradox was followed by several others: Matthew Haber (2012), Joeri Witteveen (2015), and Jerzy Brzozowski (2020). Haber argues that *Levine's Thesis* is false. Witteveen argues against Haber. Brzozowski defends Haber's position.

My argument for *C1* appeals only to biology, with no mention of theories of reference. Indeed, I take the rejection of *Levine's Thesis* to be straightforwardly present in the words of biologists themselves. So why have some of these philosophers of biology accepted *Levine's Thesis*, and all of them found the matter much more complicated? Answering that question is the other major concern of this chapter. I shall argue that discussions of *Levine's Thesis*, whether for or against, have gone awry because of mistakes about language. One mistake is about the bearing of theories of reference on the assessment of a biological claim like *Levine's Thesis*. That is the subject of conclusion *C2*, argued in section 5.4. Another mistake is about reference itself. That is the subject of conclusion *C3*, argued in section 5.5. A final mistake is about the relation between linguistic decisions and the world. That is the subject of conclusion *C4*, argued in section 5.6. In sum, the engaging debate about *Levine's Thesis* has been misguided. In section 5.7, I consider some objections.

LaPorte's *de re* reading, "Any species with a type specimen necessarily contains its type specimen", is not a major concern, but what about it? LaPorte thinks that it is false because of the possibility of the type specimen "never having been born" (2003: 587). I agree: no member is essential to a species (3.3). But, of course, he and Levine have another reason for thinking that the *de re* reading is false, one that LaPorte sets aside here (2003: 584): they reject *Essential Membership*. If no organism is essentially a member of its species, then no type specimen is. So, even if the actual type specimen for a species *is* born in another possible world, it might not be a member of that very species in that world. I reject this reasoning, of course, because I endorse *Essential Membership*. Still, I agree that no type specimen of a species is necessarily a member of that species because considerations that count against the *de dicto* reading (*Levine's Thesis*) in section 5.3 count also against the *de re* one.

Does the falsity of the *de re* reading yield another conflict, this time with the *truth* of *Essential Membership*? We shall see that it does not (5.3): although the type specimen is necessarily a member of its species it is not necessarily a member of the species for which it was selected as a type specimen.

5.2 The Causal Theory of Reference and *Levine's Thesis*

Consider Levine's path to his Thesis. It starts with David Hull's "compelling account of the role of type specimens in the practice of taxonomy" (2001: 325), an account Hull offers in urging individualism and anti-essentialism about species. Michael Ghiselin, who shares those views,[2] is led to say: "As species are individuals, there is but one rigorous way to define their names: ostensively, in a manner analogous to a christening" (1966: 209). Levine remarks: "It is interesting that Ghiselin's analogy to christening predates the literature on the Kripke-Putnam theory of reference" (2001: 336, n. 3). And Levine notes that Hull was "quick to recognize" a connection between his view of type specimens and the Kripke-Putnam theory of reference:

> the importance [Hull] ascribes to the collection of type specimens in the ostensive naming of a species is strongly reminiscent of the role played by acts of baptism or dubbing in the Kripke-Putnam theory of rigid designation. (p. 328)

Others noted this too (LaPorte 2003: 584; Haber 2012: 770; Witteveen 2015: 570; Brzozowski 2020: 2).[3]

Now I note first that it is more usual, and much better, to call the Kripke-Putnam theory of reference, the "causal" not "rigid designation" theory.[4] In any

[2] In section 2.3, I quoted Ingo Brigandt's claim that "most biologists and philosophers favor the idea that species are individuals rather than natural kinds" (2009: 77–8). I thought that Brigandt may be right about philosophers of biology, referring particularly to the debate over type specimens and reference that is the concern of this chapter. But I cited a recent survey (Pušić et al. 2017) that shows Brigandt is quite wrong about biologists.

[3] Devitt (2008, 2018a, 2018b) are among the papers cited by Brzozowski as offering "defenses of the causal-theoretical account of typification" (2020: 7). This is very odd because there is no such defense in any of these papers. Indeed, their only mention of type specimens and the causal theory of reference, in the one breath, is in a footnote sentence (2018b: 39, n. 3) that concerns something else: the sentence foreshadows the conclusion of section 5.5 below that the causal theory does not imply *Levine's Thesis.*

[4] (I) Kripke (1980) carefully defined 'rigid designator' for singular terms for the purpose of arguing that standard description theories of the reference-determining meaning of proper names are false. But, as quickly became apparent, this argument is easily avoided by modifying the description theory: a name's meaning is expressed by a *rigidified* description (Devitt and Sterelny 1999: 53–4). So, rigidity is quite compatible with some description theories. (II) The name "rigid designation" is particularly infelicitous for the Kripke-Putnam theory of "natural kind" terms. For, though Kripke extended his talk of "rigid designator" to general terms, he did not provide a definition of its use for general terms. Just what the "rigidity" of such a term amounts to, or should amount to, is unclear, as quite a large literature shows; see, for example: LaPorte (2000); Schwartz (2002); Devitt (2005).

case, what was central and most novel about the Kripke-Putnam theory was not the appeal to dubbing, which we will consider in a moment, but the idea of epistemically undemanding *reference borrowing*: people who are very ignorant, even wrong, about the referent of a term, whether a proper name or a "natural kind" term, can nonetheless be competent users of the term simply in virtue of borrowing its reference from someone who was competent; there is a causal chain of such borrowings all the way back to the people who fixed the reference in a dubbing. This was a truly revolutionary idea. And Hull embraced that too:

> In rigid designation, a name is conferred in an initial baptismal act (possibly fictitious) and thereafter passed on in a link-to-link reference preserving chain. Regardless of the appropriateness of the Kripke-Putnam analysis in general, it accurately depicts the way in which systematists introduce the names of biological taxa. (Hull 1982: 491–2)

There was nothing novel, or particularly interesting, about drawing attention to dubbings as the typical way that proper names and some "natural kind" terms get their reference. Previous theorists of reference had not failed to notice the obvious fact that the names of many entities—babies, pets, ships, newly discovered animals and substances, and so on—typically acquire reference-determining meanings at baptisms and the like. But *what* meaning and reference was thus acquired in a dubbing, and *how*? *That was the issue*. The established "description theories" all assumed that the resulting reference was determined by descriptions that all competent with the new term associated with it. The major novelty of the Kripke-Putnam causal theory was, first, to reject that theory and, second, to emphasize that the dubbers who fix the reference *then pass on the benefits of dubbings to others who may know little or nothing about the referent*. But what did the Kripke-Putnam theory tell us about that reference fixing in a dubbing? Not very much. Thus Kripke, discussing proper names in *Naming and Necessity*, talks briefly of "fixing a reference by description, or ostension" (1980: 97). Howard Wettstein thinks fixing by description was Kripke's "paradigm" (2012: 115). Putnam talks of an "ostensive definition", but one accompanied by a description (1975: 225–9): as he emphasized later, "descriptions play a key role: the original dubber or dubbers identify or have the capacity to identify what they are talking about by definite descriptions" (2001: 496–7).

Indeed, it was hard then, and is hard now, for anyone to say much about what goes on in reference fixing. Ostension always struck me as the right way to go, but then what determines that a particular object is the object of ostension? There have been description theories of that too (Reichenbach 1947; Schiffer 1978). I favored a causal theory: reference is fixed in an object, directly or indirectly, by the causal link between a person and the object when it is the focus of that person's perception. This is what I call a "grounding" (1974, 1981a).

So, on this view of reference fixing, the original users have their ability to designate Aristotle by 'Aristotle' in virtue of a certain causal link to him, and then we inherited this ability to designate him by reference borrowing. Even if one goes along with these old discussions of reference fixing, much is left unexplained, as I summarized in a recent update (2015b). Still, those discussions did include a development that is very relevant to *Levine's Thesis*, the idea of "multiple grounding". I will get to this in section 5.5.

In light of this, return to Hull and Levine on type specimens. Given their individualism, they think that the name attached to a species by a type specimen is a *proper* name (Levine 2001: 329). They clearly reject the idea that the reference of that proper name is fixed by means of a description of the Aristotelian essence of the species. But then how do they think that reference *is* fixed? Levine has this to say:

> What allows such rigid designators to attach to their referents irrespective of the truth of any associated descriptions is that *they acquire their meanings in acts of dubbing or baptism*.... The similarity between the collection of type specimens, as understood by Hull, and such acts of baptism, should be evident. In the former, a biologist, in direct contact with a part of the target species (the specimen), attaches a name to a species without thereby proposing an Aristotelian definition. (2001: 328)

The theory of grounding that I have just described is clearly a "direct-contact" view of reference fixing and so it is not surprising that Levine (2001: 330–2) is sympathetic to it (and aware of some of its difficulties).

How do we get from this sort of causal theory to *Levine's Thesis*, "Necessarily, any species with a type specimen contains its type specimen"? The Thesis comes from the following view: "No matter in which species the type specimen is placed, its name goes with it" (Hull 1982: 484). Thus, the above-quoted passage, in which Hull likens the "rigid designation" theory's treatment of the "initial baptismal act" to the introduction of "the names of biological taxa", is followed by this:

> Both...require reference preservation. The respective terms cannot change their reference, although we can find out that we are mistaken about what we thought their reference was. (p. 492)

This idea that the reference "cannot change" suggests to Levine that "the relation between a type specimen and the reference of its species name is...necessary" (2001: 334).

So Levine thinks that the causal theory applied to the species naming procedure implies *Levine's Thesis*. All his respondents agree. Now, anyone who accepts this implication and favors the causal theory might well be led to embrace *Levine's Thesis*. Indeed, that is clearly the path of Levine and LaPorte; it seems also to be the path of Witteveen, as we shall see (5.6.2). Yet is it really appropriate to embrace a biological thesis like Levine's on the basis of a theory of reference? I think not. Semantics should not be dictating to biology. Rather, semantics should answer to biology. This claim reflects the methodology of "putting metaphysics first" that I have argued for in a book of that name:

> We should approach epistemology and semantics from a metaphysical perspective rather than vice versa. We should do this because we know much more about the way the world is than we do about how we know about, or refer to, that world. (2010: 2)

It follows that it is a mistake to use *any* semantic thesis to assess *any* biological thesis; the direction of assessment should be the reverse. Applying this to our particular issue yields another one of my conclusions, *C2*: *it is a mistake to use a theory of reference to assess Levine's Thesis*. My argument for this is in section 5.4.

Still we are interested in semantics as well as biology and so we do need a theory of reference that is compatible with the biological facts including, according to *C1*, the falsity of *Levine's Thesis*. In section 5.5, I shall argue that the causal theory is compatible *once we take account of multiple grounding*; for multiple grounding allows reference to change. So, I think that Levine and his respondents are wrong to accept the above implication: *the causal theory of reference does not imply Levine's Thesis*. This is my conclusion *C3*, to be argued in section 5.5.

I turn now to an evaluation of *Levine's Thesis*, an evaluation that will, of course, make no appeal to theories of reference.

5.3 The Falsity of *Levine's Thesis*; The Case for *C1*

Haber came up with an excellent example which has appropriately been at the center of the discussions of *Levine's Thesis* and will be at the center of mine:

> In the late 1990s a minor taxonomic scuffle arose over the endangered San Francisco Garter Snake (*Thamnophis sirtalis tetrataenia*, Cope in Yarrow 1875), and the common California Red-Sided Garter Snake (*Thamnophis sirtalis infernalis*, de Blainville 1835). Researchers discovered that *T. s. infernalis*' type specimen belonged to *T. s. tetrataenia* (Boundy and Rossman 1995; Barry et al. 1996). Typically in such cases the taxa would be re-named. The codes of taxonomic nomenclature are clear on this, with rules specifying just how to handle such cases, e.g., the principles of priority and typification (ICZN 1999, Art. 23, 61). In this case, though, a petition was submitted to the International Commission on Zoological Nomenclature (ICZN) requesting that the names be conserved for each taxon in question. The case was published (Barry and

> Jennings 1998), commentary solicited (Smith 1999), and a ruling issued (ICZN 1999): Opinion 1961 of the ICZN stated that a new type specimen had been designated for *T. s. infernalis*, thus conserving prevailing usage of the names.
> (Haber 2012: 767–8)

This example is about the type specimen of a *sub*species whereas *Levine's Thesis* is explicitly about species:

> Necessarily, any species with a type specimen contains its type specimen.

Still what goes for the type specimen of a species goes for that of a subspecies. So, we should take *Levine's Thesis* as being implicitly about subspecies too. Applying it to the subspecies *T. s. infernalis*, we get:

> Necessarily, *T. s. infernalis* contains its type specimen.

Does it? The resounding answer from experts is "No". The experts we need are those who know most about the type specimens of garter snakes, biologists, particularly taxonomists. The 1835 type specimen, or holotype, for *T. s. infernalis* (originally *Coluber infernalis*) is held in a museum in Paris and catalogued as "MNHN 846" (Boundy and Rossman 1995). We shall see that some think that MNHN 846, *is not* a *T. s. infernalis* and others think that it *may well not be*. There is no sign of any expert thinking that *it must be*. So, *Levine's Thesis* is false—conclusion *C1*—and there is no paradox.

It will help to show this if we identify two propositions that are entailed by the application of *Levine's Thesis* to this example. First, and most obviously:

> **HOLO**: MNHN 846, the type specimen for *T. s. infernalis*, is an *infernalis*.

Boundy and Rossman's claimed discovery that 846 is, in fact, a San Francisco Garter Snake has not been contested. Suppose that it is so. Then, with that discovery, the application of *Levine's Thesis* entails that *T. s. infernalis* is (and always has been) the subspecies, San Francisco Garter Snake, and not, as everyone has thought for decades, the subspecies, California Red-Sided Garter Snake. For, according to the discovery, 846, the type specimen of *T. s. infernalis*, is in the former subspecies not the latter. So:

> **INF** *T. s. infernalis* is the subspecies, San Francisco Garter Snake, not the subspecies, California Red-Sided Garter Snake

The very bad news for *Levine's Thesis* is simple. There is *no sign at all* of any expert endorsing either HOLO or INF and lots of signs of their not doing so. If HOLO and INF are false then so is *Levine's Thesis*.

Consider Boundy and Rossman (1995) on HOLO. They note that a 1941 review "restricted the name *infernalis* to the California coastal subspecies" and "revived the name *T. s. tetrataenia*" for "the San Francisco Peninsula populations" (1995: 236). As a result, at the time of their paper, as other biologists remark, "the taxonomy of the western subspecies of *Thamnophis sirtalis* has been resolved and well-accepted for 45 years" (Barry et al. 1996: 172). Boundy and Rossman (1995) have a detailed discussion of whether holotype MNHN 846 should be allocated to "either of the populations currently known as *T. s. infernalis* or *T. s. tetrataenia* or of an intermediate between the two" (p. 237). They found that a certain

> combination of pattern elements on individual snakes is limited to the San Francisco Peninsula... within populations of typical *T. s. tetrataenia*. The geographic restriction of this pattern strongly indicates that the holotype of *C. infernalis* is assignable to those populations... The holotype belongs to a population(s) outside the geographic range and definition of *T. s. infernalis* as currently recognized. (p. 238)

In other words, *MNHN 846 had been misidentified and is not an infernalis: HOLO is false.*

Now consider Barry and Jennings (1998). In their petition against Boundy and Rossman's renaming proposal, they claim: "It is possible that the holotype of *T. s. infernalis* is a specimen of *T. s. tetrataenia*" (1998: 224). In other words, *MNHN 846 might have been misidentified as an infernalis and HOLO might be false. Levine's Thesis* cannot allow this because it entails that 846 *cannot* be both a type specimen for *infernalis* and not an *infernalis*.

What about INF? Boundy and Rossman reject it also, but not so obviously. First, conspicuously, Boundy and Rossman do *not* say that, given their discovery about MNHN 846, we should embrace INF. Rather, their discussion of the "allocation" of 846 proceeds as if INF is not even under consideration. Thus, in making the comparisons that the allocation requires, they examined "approximately 200 specimens from within the range of *T. s. infernalis*". And their examination leads them to say that a certain marking on *Thamnophis sirtalis* "is reduced to irregular spotting, or replaced by a broad, dark ventrolateral suffusion, in *T. s. infernalis*" (1995: 237). If INF were even a possibility given what Boundy and Rossman were revealing about 846, then rather than talk simply, as they do, of "*T. s. infernalis*", they should have said something like "the coastal snakes that *may have been wrongly identified as T. s. infernalis*". They are taking the falsity of INF for granted.

It's a similar story with Barry and Jennings (1998). As noted, they accept the possibility that MNHN 846 is a *tetrataenia*. If *Levine's Thesis* were right, then this possibility would entail the possibility that INF is true. Barry and Jennings write as if this possibility has never occurred to them; Smith (1999), likewise. Thus, Barry and Jennings, after citing a large range of literature describing the San Francisco Peninsula snakes as "*T. s. tetrataenia*", claim that "much of the same literature refers to *T. s. infernalis* as an allopatric form that does not occur on the San Francisco Peninsula" (1998: 225–6). There is no airing of the idea that this literature might be wrong because, given the facts about MNHN 846, *infernalis* might be *tetrataenia* and so INF might be true. Rather, Barry and Jennings presume INF is false.

Boundy and Rossman's discovery about MNHN 846 does not even raise the issue, for taxonomists, of whether the coastal snakes are *T. s. infernalis.* The issue actually raised by the discovery is quite different and is indicated by Haber: "typically in such cases the taxa would be re-named" (2012: 768). The issue raised is simply *which official names to use for the subspecies of Thamnophis sirtalis in the future.* Nothing more, nothing less. Should taxonomists follow the "default" (Haber 2012: 777), according to the ICZN code, *re*naming *tetrataenia* "*infernalis*" and assigning a new name to *infernalis*, as Boundy and Rossman propose? Or should both subspecies retain their old names, as Barry and Jennings successfully petitioned? *All* parties see the issue raised by the discovery as simply over future names. Thus, for Boundy and Rossman, it is an issue of "nomenclatural changes" (1995: 238); for Barry and Jennings, one of "the rearrangement of the subspecies names" (1998: 226); for commentator Smith, one of "the stability of usage of these names" (1998: 72); finally, for the Commission, ICZN itself, in opinion 1961, the issue is

> the conservation of the subspecific name of *Thamnophis sirtalis infernalis* (Blainville, 1835) for the California red-sided garter snake from the Californian coast, and of *T. s. tetrataenia* (Cope in Yarrow, 1875) for the San Francisco garter snake from the San Francisco Peninsula ... (2000: 191)

This common understanding of the issue raised by MNHN 846 is at odds with INF and hence with *Levine's Thesis.* For, if INF were correct, there could be no question of *conserving* '*T. s. infernalis*' for the coastal snake since it would already be the name for the Peninsula snake not the coastal snake. And there could be no question of *re*naming the Peninsula subspecies '*T. s. infernalis*' because it would already have that name (even though nobody realized that it had!). It would have that name because MNHN 846 is the type specimen for *T. s. infernalis* and 846 is a Peninsula snake. The possibility that INF might be true is not even contemplated.

What then should we conclude from this discussion of Boundy and Rossman's uncontested discoveries about the type specimen, MNHN 846? We should conclude that those who know most about garter snakes think that HOLO and INF, two theses about garter snakes, are, or at least may be, false. Taxonomy is rife with controversies but this is not one of them. So, the biological experts reject *Levine's Thesis.* So we should too: conclusion C1.

I noted in section 5.1 that *Levine's Thesis* is LaPorte's *de dicto* reading of a claim that also has the following *de re* reading: "Any species with a type specimen necessarily contains its type specimen" (2003: 586). This reading is not a main concern but it is worth noting that the present discussion counts against that reading too. MNHN 846 was the type specimen for *T. s. infernalis*. Boundy and Rossman's uncontested discovery was that 846 had been misidentified and was not an *infernalis*. So the *de re* reading is false.

It is important to see that the falsity of the *de re* reading poses no problem for *Essential Membership* (Chapter 4). According to *Essential Membership*, 846 is necessarily a member of its subspecies, *T. s. tetrataenia*. That is of course consistent with 846 being not necessarily a member of the subspecies for which it was the type specimen, *T. s. infernalis*, indeed with it not being a member at all. So, the falsity of the *de re* reading does not create a new paradox.

5.4 "But What about the Theory of Reference?"; The Case for *C2*

In section 5.2, I foreshadowed the conclusion *C2*, that "it is a mistake to use a theory of reference to assess *Levine's Thesis*". Rather, the direction of assessment should be from biological facts to the theory of reference. So, my discussion of HOLO and INF has proceeded without appeal to a theory of reference. But *why* is it a mistake to make such an appeal? Why should we not follow Levine and others and argue as follows? "Our favorite theory of reference for biological kind terms, *TR*, tells us that, given the nature of MNHN 846, the name '*T. s. infernalis*' refers to the Peninsula snakes not the coastal snakes. So HOLO, INF, and *Levine's Thesis*, are true after all!" Problem: *Why believe TR*? Why not prefer a rival theory that tells us that '*T. s. infernalis*' refers to the coastal snake, or even to nothing at all? The traditional answer has been that *TR is supported by our referential intuitions*. Thus, let's suppose, *TR* predicts, time and again, that the reference of a biological kind term *E* in real or imagined situations is *X* and it just seems intuitively to us philosophers that *E* does indeed refer to *X*. That's the methodology of the theory of reference. But this methodology has been severely criticized in recent years. Many have argued that the methodology is scientifically unsound and have insisted that theories of reference must be tested experimentally; see, for example, Machery et al. (2004); Machery et al. (2009); Nichols et al. (2016). Genoveva Martí (2009, 2012, 2014) and I (2011b, 2012a, 2012b, 2015a) have joined in the criticism and have gone on to argue that theories should be tested against *linguistic usage*.

This debate over methodology cannot of course be replayed here,[5] but I shall briefly apply the Martí-Devitt line to the present example. We should not accept any theory of reference for a term simply because its predictions conform to our intuitions about what the term refers to. Rather, we should test the theory against

[5] See Devitt and Porter (2021) for a summary of the literature and some examples of testing usage.

the usage of those competent with the term. So, if *TR* is to be applied to biological terms, it needs to be tested against the usage of biologists. Do these people *show by their usage* that they are referring to biological *X* by term *E*? For example, does the taxonomists' use of '*T. s. infernalis*' show that they identify the Peninsula snake as its referent? Moral: *we need biologists' opinion on the likes of INF in order to know whether TR is right*. Our only way now, perhaps ever, to determine scientifically whether a theory of reference for biological terms is right depends on our determination of the biological facts. The biologists' usage shows us that INF is false, as we have seen. So *TR* is false. That is the right direction of argument. No theory of reference has the evidential support to rule on INF and *Levine's Thesis*, contrary to what Levine and others presume.

That is my present argument for *C2*. Further support for *C2* can be found in my *Putting Metaphysics First* (2010).

Though a theory of reference should not be used to assess *Levine's Thesis*, any such theory should be able to explain the linguistic usage demonstrated here, as anywhere. The causal theory mentioned in section 5.2, unlike *TR*, does explain that usage, once developed to include "multiple grounding".

5.5 The Causal Theory of Multiple Grounding; The Case for *C3*

As noted, my theory of "grounding" is a theory of the sort of reference fixing by "direct contact" that Hull and Levine favor. The most obvious examples of such groundings are the ceremonial dubbings that they mention. But there can be groundings without any such dubbings. Thus, consider the naming of the cat Nana, discussed by Levine (2001: 330–1). This naming was by a dubbing but it could have been simply the result of usage: someone looking at Nana might have just said "Nana is a striking looking kitten" and thereby started the practice of calling the kitten "Nana". Nicknames are often introduced in this way. I recently summed up the theory of grounding as follows:

> What is it about all these situations that ground the name in a certain object? It is the causal-perceptual link between the first users of the name and the object named. What made it the case that this particular object got named in such a situation was its unique place in the causal nexus in the grounding situation. (2015b: 114)

This leads straightforwardly to the theory of *multiple* grounding.

> It is important to note that this sort of situation will typically arise many times in the history of an object after it has been initially named: names are typically

> *multiply grounded* in their bearers. These other situations are ones where the name is used as a result of a direct perceptual confrontation with its bearer. The social ceremony of introduction provides the most obvious examples: someone says, "This is Nana", demonstrating the kitten in question. Remarks prompted by observation of an object provide many others: thus, observing Nana's behavior, someone says, "Nana is skittish tonight". Such remarks are likely to happen countless times during Nana's life. All these uses of a name ground it in its bearer just as effectively as does a dubbing because they involve just the same reference-fixing causal-perceptual links between name and bearer. . . . Dubbings and other first uses of a name do not bear all the burden of linking a name to the world. (p. 114)

I used this idea of multiple grounding, together with Hartry Field's (1973) idea of partial reference, to explain cases of reference *confusion* (1974: 200–3). Thus, consider Kripke's famous leaf-raking example: "Two people see Smith in the distance and mistake him for Jones" (1979: 14). Suppose one person comments to the other, "Jones is raking the leaves". I argued that this use of 'Jones' has a semantic-referent, Jones, but no determinate speaker-referent; both Jones and Smith are *partial* speaker-referents because the use is grounded in both (1981b: 512–16; 2015b: 118–21). Later (1981a: 138–52; 2015b: 121–4), I applied the ideas to cases of reference *change* including another famous example, Gareth Evans' 'Madagascar' (1973). The story goes that Marco Polo, on the basis of a hearsay report of Malay sailors, mistakenly took the name of a portion of the African mainland, 'Madagascar', as the name of the great African island. And that island is now, of course, the semantic-referent of 'Madagascar'. So 'Madagascar' changed its reference. The explanation, in brief, is that the reference of a name changes from *x* to *y* when *the pattern of its groundings changes* from being in *x* to being in *y*.[6] This discussion is particularly relevant to *Levine's Thesis* if we go along with the individualist view that a species name is a proper name.[7]

Appeal to multiple grounding is also vital in explaining reference change in "natural kind" terms (Devitt 1981a: 190–5). Arthur Fine (1975: 22–6) criticized Putnam's causal theory of these terms on the ground that it makes it impossible for a term to change its reference: its reference is fixed by the original dubbing. Yet

[6] Nonetheless, the mistaken idea that cases of reference change are "decisive against the Causal Theory of Names" (Evans 1973: 195) persists (Searle 1983; Sullivan 2010; Dickie 2011). Kripke's own response to 'Madagascar' is in "Addenda" to *Naming and Necessity* (1980: 163). As I note (2015b: 123, n. 33), the grounding theory can be seen as an explanation of Kripke's admittedly brief proposal (but doubtless not one he would accept).

[7] So, it is odd that Levine does not mention this theory of reference change. He devotes much attention (2001: 330–2) to a discussion of "the qua problem" in Chapter 4 of Devitt and Sterelny (1999), a textbook presentation of the causal theory of reference. That presentation includes the theory of reference change (pp. 75–6). Indeed, in the 1987 first edition which Levine uses, the theory of reference change immediately precedes the discussion of the qua problem.

such scientific terms quite obviously often do change their reference. I pointed out (1981a: 291–2, n. 1) that Putnam could easily add multiple grounding to his theory. And later he did: "As Devitt rightly observes, such terms are typically 'multiply grounded'" (2001: 497). Reference change can then be explained, as it was with proper names, as a *change in the pattern of groundings* (Devitt 1981a: 192–5). This discussion would be particularly relevant to *Levine's Thesis* if we do not accept individualism as, it seems, most biologists do not (2.3).

This explanation of reference change is not an ad hoc addition to the causal theory to solve problems. It is a straightforward corollary of the causal theory of groundings:

> Groundings fix designation. From the causal-perceptual account of groundings we get the likelihood of multiple groundings. From multiple groundings we get the possibility of confusion through misidentification. From confusion we get the possibility of designation change through change in the pattern of groundings.
> (Devitt 2015b: 123–4)

It is a truism among theorists of language that an expression gets its meaning and reference from conventions of usage. These conventions sometimes start with stipulations—dubbings are examples—but they mostly come from regular usage. *However* a convention is established, *even if by stipulation*, it can change through regular usage. (Think of the sad fate of 'beg the question'.) The above theory of groundings is an explanation of change for some sorts of words.[8]

We now apply this theory to the names used to refer to Haber's garter snakes. An expression's conventional reference is typically established by regular usage. There was clear consensus among taxonomists in the above debate that since 1951 there had been a stable usage of the name '*Thamnophis sirtalis infernalis*' to refer to California Red-Sided Garter Snakes; see Barry and Jennings (1998), particularly. According to the causal theory this stability reflects *a pattern of groundings* of the name in those coastal snakes, a pattern of taxonomists (and others) using the name as a direct result of perceptual contact with those snakes. Doubtless in those decades, there were some groundings of the name in snakes of other kinds, particularly in MNHN 846, which is, after all, the type specimen for *T. s. infernalis* and yet is (we are assuming) a *tetrataenia*, not an *infernalis*. But these misidentifications pale into insignificance against the pattern of groundings in the coastal snake, *infernalis*. That pattern established and maintained the conventional use of the name '*Thamnophis sirtalis infernalis*' to refer to the coastal snake. And this is true whether we take the name to refer to an individual or to snakes of a certain kind.

[8] For a discussion of the role of conventions in explaining meanings, see Devitt (2021b: 75–93).

According to Article 61 of the code, MNHN 846 should have provided "the objective standard of reference" (ICZN 1999) for '*Thamnophis sirtalis infernalis*': type specimens are supposed to stipulate a conventional usage. That is the thought behind Witteveen's claim: "If we baptize a specimen that belongs to some taxon as name-bearer, we thereby fix the name's reference to the taxon the specimen belongs to" (2015: 581). But the reference is thereby fixed only if all goes well for the stipulation. For, as just noted, stipulations can fail because expressions are not used as stipulated and different conventions are established.[9] The consensus opinion about the usage of '*Thamnophis sirtalis infernalis*' shows that MNHN 846 is an example of such failure.

I emphasize that the Hullian idea that reference "cannot change" was *never* part of the Kripke-Putnam causal theory. Certainly the issue of reference change was not addressed in the earliest presentations of the theory. Still it was in later ones.

That is the case for *C3*: the causal theory of reference does not imply *Levine's Thesis*, as Levine and others think.

C2 identified the mistake by Levine and others of using a theory of reference to determine a biological thesis (5.4). That mistake is compounded by using a theory that does not accommodate reference change.

C3's rejection of the inference from the causal theory to *Levine's Thesis* has consequences for what Haber and Brzozowski say about reference. Given their acceptance of the inference, they take their arguments against *Levine's Thesis* to count against the causal theory (semantics appropriately answering to biology; (5.2)).[10] Thus, Haber thinks that his argument "suggests that rigid designation and causal theory of reference may be more fragile than supposed" (2012: 768).[11] The argument presents "a serious challenge to philosophical accounts of proper names, or perhaps their applicability to biological taxonomy" (Haber 2012: 781). Brzozowski is led to the view that taxon names have their reference fixed by descriptions and are "descriptive names". He thinks that this "account of taxon names is able to better account for the uses and misuses of taxon names when compared to the causal view" (2020: 23). *C3* undermines these criticisms of the causal theory.

[9] A corollary is that the following claims are false: "taxonomists had always known (with a priori certainty) that the *infernalis* type specimen belonged to the *infernalis* taxon" (Witteveen 2015: 582); "Type specimens... can be known a priori to belong to [their respective species]" (LaPorte 2003: 583). Knowledge of referential facts, indeed knowledge of semantic facts in general, is always empirical (Devitt 2011a; Salmon 2020).

[10] If the rejection of *Levine's Thesis* poses a problem for the causal theory then, as LaPorte points out, it is "a general one": "it arises whether species are individuals or kinds, given the standard causal theory of reference" (2003: 586).

[11] Haber adds the following startlingly false claim: "Taxonomic theory is, in part, a theory of reference applied to biological nomenclature" (p. 768). Taxonomic theory does specify *a practice for the stipulation of* a taxon name that will *cause* it to have a certain reference when all goes well, which it sometimes doesn't; but taxonomic theory is far from a *theory of* this reference.

5.6 Philosophical Evaluations of *Levine's Thesis*

I turn now to the evaluation of *Levine's Thesis* by other philosophers. These evaluations include some claims which, from the perspective I have presented, are dead right. But they include others that are dead wrong. Thus, on the right side, Haber claims, contrary to HOLO, that "researchers discovered that *T. s. infernalis'* type specimen belonged to *T. s. tetrataenia*" (2012: 768) and goes on to reject *Levine's Thesis* and hence resolve the paradox. Brzozowski makes a similar claim (2020: 10) and endorses Haber's rejection. Even Witteveen, who wrongly endorses *Levine's Thesis*, nonetheless apparently rejects INF in saying that Boundy and Rossman "discovered that taxonomists had been wrong about which taxon was [the *infernalis* type specimen's] taxon" (2015: 582).

But then there is the wrong side.

5.6.1 Haber; The Case for C4

Haber's rejection of *Levine's Thesis* is strangely qualified: he thinks that the Thesis "only holds under idealized conditions" (2012: 782). This reflects a more serious problem: his reason for rejecting the Thesis confuses changing language with changing the world. This is the last of the "mistakes about language" that are a major concern of this chapter.

My own reasons for rejecting *Levine's Thesis* arose from two related responses of taxonomists to the discovery about MNHN 846, the type specimen for the subspecies *T. s. infernalis*. These responses were contrary to what the Thesis demands. First, contrary to HOLO, these experts concluded that 846 had been, or might have been, misidentified as an *infernalis*, the California Red-Sided Garter Snake; second, contrary to INF, these experts showed no sign of even entertaining the possibility that *infernalis* was not that coastal snake.

Now as noted in section 5.3, the discovery about MNHN 846 did demand a further response: taxonomists, particularly ICZN, had to make a decision about the future official names for the subspecies of *Thamnophis sirtalis*. *But the falsity of Levine's Thesis does not depend in any way on that decision about future usage*. Yet, as we shall see, Haber seems to think that it does. He seems to think that the Thesis *would be* true if ICZN always followed the code's "default" in such cases of misidentification, a default that would have been illustrated had ICZN accepted Boundy and Rossman's proposal that *tetrataenia* be renamed "*infernalis*" and a new name be assigned to *infernalis*.

Abraham Lincoln is said to have once pointed out that a person's calling a donkey's tail a "leg" does not make it a leg. Similarly, the ICZN's calling the Peninsula snake "*T. s. infernalis*" would not have made it *T. s. infernalis*. It was a worldly fact that the Peninsula snake was not *T. s. infernalis*, no matter what decisions ICZN, or anyone, makes about how to use language in the future. Contrary to what postmodernists, and sadly many others, seem to think, languages do not make worlds. This is not the place to argue this large issue (but see, for example, Devitt 1997: 235–58; 2010: 99–136).

The key discussion in Haber begins nicely:

> That a specimen was preserved and identified prior to careful study of a particular taxon does not mitigate that the type specimen may be wrongly hypothesized to belong to that taxon. (2012: 779)

But then Haber goes on:

> In a default case, the species identity of the type specimen does not change, it still belongs to the species it designates. (p. 779)

Had ICZN responded to the discoveries about MNHN 846 by deciding to follow the default it would have renamed *tetrataenia* "*infernalis*". This would have changed the status of 846: before such a decision, 846 does not belong to the subspecies for which it was a type specimen because it does not belong to *infernalis*; after the decision, it would have belonged to the subspecies for which it was a type specimen because it belongs to *tetrataenia*. But it would not have been in virtue of this decision that 846 kept its "species identity"! 846 was a *tetrataenia* (we are assuming) misidentified as an *infernalis*, showing *Levine's Thesis* to be false, *whatever linguistic decision anyone made about future usage*. Haber continues:

> On successful active petition... the type specimen... is reassigned to a new species, and no longer belongs to the species it formerly designated (though other specimens might). (p. 779)

As Witteveen points out, Haber is arguing that the decision by ICZN to accept the petition of Barry and Jennings "entails that a type specimen got misidentified" (Witteveen 2015: 575). Yet, what ICZN actually did was decide to conserve the subspecific *names* of both *T. S. infernalis* and *T. s. tetrataenia* (2000: 191), rather than follow the default. This decision did not reassign MNHN 846 "to a new species" or entail that 846 had been misidentified. On the contrary, the decision is totally irrelevant to what (sub)species 846 belongs to. 846 had been misidentified as an *infernalis*, independent of any linguistic decision: to repeat, languages don't make worlds. Finally, contrary to what Haber claims (2012: 780), it is not because of that decision, rather than the default one, that the "*de dicto* necessity [*Levine's Thesis*] fails to hold". It fails simply because type specimens can be misidentified, as 846 illustrates. The "species identity" of any type specimen, like that of any organism, is constituted by its nature not by a linguistic decision of ICZN.

In sum, *it is a mistake to make any inferences about species identity, and hence about Levine's Thesis, from decisions about nomenclature.* This is my conclusion *C4.*

5.6.2 Witteveen

Witteveen claims to resolve Levine's paradox by arguing that "there is no sense in which type specimens belong contingently to the species they name" (2015: 571). Well, if my argument against *Levine's Thesis* is right then there is at least one such sense. Set that aside for a minute. According to LaPorte, there is another sense: the contingency that arises from the rejection of the *de re* necessity, "Any species with a type specimen necessarily contains its type specimen". I argued that the misidentification of MNHN 846 provides one reason against this necessity (5.3). And LaPorte rightly points out that we should reject the necessity because of the possibility of the type specimen "never having been born" (2003: 587). Furthermore, he thinks, though I do not (5.1), that we should also reject this necessity because *Essential Membership* is false. So, there are *several* potential reasons for the contingency that comes from rejecting LaPorte's *de re* necessity. How does Witteveen resist all of them in claiming that "there is no sense in which type specimens belong contingently to the species they name"? Briefly, by confusing LaPorte's *de re* reading with his *de dicto* one (in a section called "Contingency confusion"!):

> Thus, it appears that in all possible worlds in which we find a species with a type specimen, it contains its type specimen. This means that the sentence "Any species with a type specimen necessarily contains its type specimen" is true after all. (Witteveen 2015: 576–7)

This is wrong. What appears to Witteveen to be so in his first sentence amounts to, "Necessarily any species with a type specimen contains its type specimen". This is LaPorte's *de dicto* reading, *Levine's Thesis*. This differs strikingly in the scope of its 'necessarily' from what Witteveen takes the sentence to mean in his second sentence, namely, LaPorte's *de re* reading. And, the contingency we are considering is a rejection of the *de re* reading *not* the *de dicto* one. Witteveen has not addressed *that* "sense in which type specimens belong contingently to the species they name".

Return to Laporte's *de dicto* reading, *Levine's Thesis*. Witteveen's endorsement of this is, for our purposes, the key sense of contingency that he rejects. So, what is Witteveen's case for *Levine's Thesis*? It starts with criticism of Haber's case against. We have just rejected Haber's argument that the ICZN decision to accept Barry and Jennings' petition establishes that MNHN 846 was misidentified. Witteveen's criticisms are different. First, he claims:

> What Haber should have said is that [that ICZN decision] causes a specimen *that formerly served as type specimen* to stop belonging to the taxon for which it formerly anchored the taxon name. (2015: 580)

Now that decision *did* cause MNHN 846 to cease to be the type specimen of *infernalis*. But the decision *did not* cause 846 "to stop belonging to" *infernalis*: 846 never did belong. And no decision by ICZN could bear on the worldly fact of 846's subspecies membership; see conclusion *C4*. Witteveen's second criticism is better: he claims that the ICZN decision "does not show that *de dicto* necessity [*Levine's Thesis*] fails" (p. 581). No linguistic decision *could* show this. So Witteveen is right that Haber's case *against Levine's Thesis* fails. But what does Witteveen have to say *for Levine's Thesis*? Only the passage we quoted and rejected earlier (5.5): "If we baptize a specimen that belongs to some taxon as name-bearer, we thereby fix the name's reference to the taxon the specimen belongs to" (p. 581). The problem was that attempts to stipulate usage can fail; reference can change (5.5). In any case, no thesis about language has the authority to settle a biological matter; see conclusion *C2*. To support *Levine's Thesis*, Witteveen needs to show that MNHN 846 was *not* misidentified as an *infernalis*, as taxonomists clearly think it (very likely) was. Witteveen has not done so.

5.6.3 Brzozowski

Brzozowski offers "a defense of Haber's (2012) position in response to Witteveen (2015)" (2020: 4). Part of this defense is the rejection (2020: 12) of a criticism of Haber that I have just emphatically endorsed: the criticism that Haber takes *the ICZN decision* to entail that a type specimen got misidentified. In rejecting this criticism, Brzozowski points to a passage (Haber 2012: 778) like the one above that I labeled "on the right side". But the criticism is well-based in the cited passages "on the wrong side".

Brzozowski's discussion of this criticism, and his own remarks "on the right side" (2020: 10), might suggest that he rightly thinks that the biological discovery that MNHN 846 had been misidentified alone shows that *Levine's Thesis* is false. But, in fact, he thinks that this discovery falsifies only a "metalinguistic" version of the thesis about "the reference of a species name" (p. 22). And this falsification depends on complicated semantic machinery, including the claim that names are descriptive (pp. 14–23). This is a mistake: biology alone shows *Levine's Thesis* false. No semantics is needed; see conclusion *C2*.

I turn finally to some likely objections to my argument against *Levine's Thesis*.

5.7 Objections

I have a good basis for anticipating objections. For, the argument in this chapter has been presented before in a paper, "Type Specimens and Reference", that was rejected by two journals on the basis of some thoughtful reports from reviewers.[12] I found the objections from two of these reviewers particularly interesting. The reviewers rightly think that issues about language have been center stage in the discussion of *Levine's Thesis* and *they insist that these issues continue to be*. Indeed, they find it incomprehensible that linguistic issues should not be put center stage. So, the reviewers are insisting on precisely what my paper argues is a very mistaken methodology. I shall develop my argument in this section in responding to the objections. It seems that this linguistic methodology is much more entrenched in this area of the philosophy of biology than I had supposed.

5.7.1 Reviewer R1 and Codes of Nomenclature

The objections from *R1* do not seem to be about language to begin with. *R1* claims that my

> bold argument would have been very interesting if it had been supported by convincing empirical evidence that taxonomists agree unanimously that it is not necessary for type specimens to belong to their species... I expected that the author would present evidence from questionnaires with vignettes of the kind that are frequently encountered in contemporary experimental philosophy (particularly in the area of semantics).

Section 5.3 presents fairly overwhelming evidence that *all* the taxonomists involved in the case of MNHN 846, and the international body ICZN itself, agree that 846, which is indubitably the type specimen for *Thamnophis sirtalis infernalis*, is, or at least might be, nonetheless a *T. s. tetrataenia*. What they agree on is inconsistent with *Levine's Thesis*. Now it is always good to have more evidence. So, we could see what taxonomists say about other cases of apparently misidentified type specimens. And we could indeed do some "experimental philosophy" on taxonomists. But if we do, we should not ask the taxonomists their opinion about whether it is "necessary for type specimens to belong to their species" (*Levine's Thesis*): that sort of question asked of taxonomists is far too abstract and "philosophical" to provide good evidence. Rather, we should ask taxonomists about actual or imagined cases of apparently misidentified type

[12] The journals were *Biology and Philosophy* and *History and Philosophy of the Life Sciences*.

specimens. This would provide good and direct evidence for or against *Levine's Thesis* of just the same sort as I provided. Indeed, we could present taxonomists with a vignette about MNHN 846 itself and ask them whether it is a *T. s. infernalis* or a *T. s. tetrataenia*; we could ask them about HOLO. But do we really need any of this extra evidence? Thus, given the *actual* discussion of 846 that I cited, we can surely be confident about the taxonomists' answer: 846, the type specimen for *T. s. infernalis* is, or at least might be, a *T. s. tetrataenia*.

This can't be *R1*'s real worry about evidence and it soon becomes apparent that it isn't. The real worry is that the evidence that I provide from that actual discussion is "not viewed in the context of the debate" of Haber, Witteveen, and Brzozowski. What context is that? *A context that is largely about language*. Thus *R1* demands

> a close analysis of how this [rejection of *Levine's Thesis*] is supported by the wording of codes of nomenclature (ICZN, ICN and others) that taxonomists have devised and follow in their nomenclatural practices.

R1 charges that I do not "attend to the role of codes of nomenclature in taxonomic practice". *R1* finds this

> really quite baffling, since these codes – and their role in taxonomic practice – have been at the center of discussion in recent contributions to the "type specimen debate". By failing to consider the content and application of the codes in taxonomic practice, the author misses entirely what this type specimen debate has been about.

R1 is, of course, right that the debate over *Levine's Thesis* has centered on such linguistic matters. Indeed, I emphasized this at the very beginning of my discussion. So, I haven't *missed* it. Rather, I have emphatically *rejected* it: a "major concern" of the paper, and this chapter, is to argue that the debate has "gone awry because of mistakes about language" (5.1).

How *might* a nomenclatural practice bear on *Levine's Thesis*? Here's a way. In section 5.4, I noted that a theory of reference, *TR*, could be brought to bear by telling us that, "given the nature of MNHN 846, the name '*T. s. infernalis*' refers to the Peninsula snakes not the coastal snakes", thus supporting *Levine's Thesis*. Now suppose that *TR* tells us this about the name '*T. s. infernalis*' *because TR* takes the nomenclatural practice of stipulating a meaning for a taxon name via a type specimen to be what constitutes that reference to the Peninsula snakes. Then, clearly, the nomenclatural practice would provide evidence for *Levine's Thesis*. But, also clearly, the practice does so only if *TR* is *right* to give this role to the practice. And the problem is that *TR* is *not right* to. How do we know? Well, for '*T. s. infernalis*' to refer to the Peninsula snakes, there would have to be a convention of using it to so refer. That's a truism. And the usage by biologists shows that there is no such convention. Indeed, biologists had for decades been identifying the *coastal* snakes, not the Peninsula ones, as *T. s. infernalis. It is these identifications by biologists that provide the evidence for or against any theory of reference of 'T. s. infernalis'* (Devitt and Porter 2021). Those identifications are what *TR* has to be tested against, and it fails.

But the moral of this tale is deeper. To assess *Levine's Thesis*, we need to know whether MNHN 846, the type specimen for *T. s. infernalis*, is a *T. s. infernalis* (HOLO). The deep moral is that it was a mistake to bring a theory of reference to bear on this question from the start (5.4). For, any theory of the reference of '*T. s. infernalis*' has to be tested against the term's usage. And the usage in question is that of taxonomists in identifying snakes as *T. s. infernalis* or not. So, to assess *Levine's Thesis*, we should simply check what biologists *do* identify as *T. s. infernalis* or not and skip the detour into the theory of reference. And that is what I did in section 5.3.

No application of a nomenclatural code constitutes the reference of 'T. s. infernalis'. That's a fact from the theory of language. There is no call for *R1* to be baffled by my inattention "to the role of codes of nomenclature in taxonomic practice". I attend to the only role played by these codes that is *relevant* to the reference of '*T. s. infernalis*'. That role, I argue (5.5), is a *causal* not constitutive one. The application of a code is an obvious attempt to *stipulate* a term's reference, for important scientific purposes. And, of course, those attempts are mostly successful: they establish a convention, thus causing the term to *have* that very reference. But, as the case of '*T. s. infernalis*' shows, sometimes stipulations fail because usage establishes different conventions. In sum, when all goes well for an authoritative body like ICZN, its stipulation that *E* is to refer to *S* will cause *E* to refer to *S*, but it never constitutes it so referring. That *E* refers to *S* is constituted by dispositions among *E*'s users (Devitt 2021b: 75–81).

Despite the irrelevance of theories of reference to the assessment of *Levine's Thesis*, we do of course need a theory of reference that is compatible with the biological facts of the matter. I offered a causal theory of multiple grounding (5.5). *R1* is not impressed, accusing me of failing "to see that taxonomists have agreed on the convention that only type designations 'ground' formal taxonomic names". Not guilty! Rather, what *R1* has failed to see is that *conventions agreed on may not be followed*; Geneva Conventions provide one example; '*T. s. infernalis*', another. *R1* continues: "One could in fact argue that one of the main purposes of the type method is to formally forbid 'multiple groundings' of taxon names." One could, but multiple groundings are a fact of linguistic life. So, it would be more plausible to argue that "one of the main purposes of the type method is to formally forbid" *groundings in any organism that is not in the same taxon as the type specimen.* That's plausible because the type method is a stipulation and stipulations indicate what people want. But, sadly, wanting something to be so, doesn't make it so.

Thus, despite the Geneva Conventions, people got tortured. Similarly, despite the ICZN code, '*T. s. infernalis*' got multiply grounded in the coastal snake. So, the term *actually* refers to that snake. And *actual* reference matters to the theory of reference, not what the ICZN, or anyone, wants.

One might put my main point in response to *R1* as follows. The empirical methodology for the theory of reference, discussed in detail in the many works cited in section 5.4, and briefly described in that section and above, shows that the linguistic "context of the debate" over *Levine's Thesis* is mistaken. *R1* insists on that context without any recognition of that empirical methodology.

5.7.2 Reviewer R2 and the Linguistic Turn

R2 characterizes my methodology as follows: "we should simply ask experts (i.e. taxonomists) about whether *Levine's Thesis* holds." That's not quite right. My refutation of *Levine's Thesis* rests entirely on what taxonomists had to say about certain snakes, organisms that taxonomists know a lot about. The refutation does not rest at all on what taxonomists think about *Levine's Thesis*, a philosophical thesis that they might well find quite puzzling. In any case, *R2* objects:

> This methodology needs further motivation, since it is far from clear... that the taxonomists actually draw the conclusion that the Author claims they do. In particular, the Author will need to consider that the taxonomists he cites recognize the difference between the *usage* of names and their *valid* designation.... it is not evident that the taxonomists think that the valid name for a taxon can refer to a taxon that doesn't include the type for that name.... the Author appears to be holding the taxonomists to unreasonably high philosophical standards of precision in talking about naming and reference.... We can't expect taxonomists to neatly distinguish between these kinds of reference in their writings.

The opinions of taxonomists about snakes that I cite, including about type specimen MNHN 846, are inconsistent with *Levine's Thesis*. That is why we should reject *Levine's Thesis*. *R2* objects that we shouldn't reject it until we know what taxonomists think about *the names* of those snakes, until we have established that taxonomists have certain quite subtle *semantic* views. But, I responded to *R2*'s review, it was a central theme of my paper that views about language should *not* be used to assess a biological thesis like *Levine's Thesis*; see *C2* (5.4). Any views about language, even ones held by expert semanticists, let alone by taxonomists, should not count against the views of expert taxonomists *about organisms*.

R2 was hugely unimpressed with this response, insisting that semantics *must* play a role. In particular *R2* finds it "really quite puzzling" how I "*could* think" that *Levine's Thesis* "is a purely biological thesis". For,

> a type specimen (a holotype or neotype) is nothing other than a specimen that serves as the bearer of a species name. So, we could rewrite [*Levine's Thesis*] as: "Necessarily, any species with a specimen that serves as the bearer of that species' name contains that specimen".[13] Is this a "purely biological" thesis? Surely not! It has semantics written all over it! Just consider a simple question this thesis invites: which is the species that the name-bearing specimen belongs to? Is it the name's semantic referent?

A consequence of *C2* is that this move to a semantic question is uncalled for and mistaken. Take our case of MNHN 846. Everyone agrees that 846 is the type specimen that serves as the bearer of the name for the species *T. s. infernalis*. Then *R2*'s "simple question", applied to this case, is: "Does MNHN 846 belong to the semantic referent of '*T. s. infernalis*'?" But the question that should concern *Levine's Thesis* is not this partly semantic one but rather the entirely nonsemantic, "Is MNHN 846 a *T. s. infernalis*?" (cf. HOLO). And the resounding answer from people who know a lot about snakes, but probably very little about semantics, is "No (or probably not)". *That* is the judgment that refutes *Levine's Thesis*. *R2*'s insistence on bringing in semantics (without even addressing my argument that we should not) is very revealing of just how entrenched this "linguistic turn" is in this area of the philosophy of biology.

There is no sign that biologists involved in this case ever entertain *Levine's Thesis*, but they show by their practices that they reject it. So, they are not bothered by the problem allegedly posed by the Thesis. And they are right not to be. The alleged problem is a philosophical illusion, a misguided attempt by philosophers, driven by mistaken ideas about the relevance of views about language, to impose a problem on biology.

5.8 Conclusion

Levine (2001) sees a conflict between the contingency of species membership and a view of the role of type specimens that he takes from Hull: "*qua organism*, the type specimen belongs to its respective species contingently, while *qua type*

[13] *R2* actually proposed the following rewrite: "Necessarily, any species with a specimen that serves as the bearer of a species name belongs to the species of which it bears the name." But this must be a slip as it is clearly not a rewrite of *Levine's Thesis*. I have made corresponding adjustments in what follows the slip.

specimen, it belongs necessarily"; he finds this "paradoxical" (p. 334). My concern has been with the thesis about type specimens which, following LaPorte, I take to be the *de dicto* necessity, "Necessarily, any species with a type specimen contains its type specimen" (2003: 586). I called this "*Levine's Thesis*". I have used Haber's lovely example of MNHN 846, the type specimen for *Thamnophis sirtalis infernalis*, to argue for conclusion *C1*: *Levine's Thesis* is false (5.3). For, the uncontested discovery by two taxonomists, Boundy and Rossman (1995), is that 846 is not a *T. s. infernalis* but a *T. s. tetrataenia*.

The alleged paradox has led to papers not only from LaPorte but also from Haber (2012), Witteveen (2015), and Brzozowski (2020). My argument for *C1* appealed only to biology, with no mention of theories of language. In this respect it differs from other arguments about *Levine's Thesis*, whether for it or against it. A major concern of this chapter has been to show that these arguments have gone awry because of mistakes about language.

First, Levine's path to *Levine's Thesis* rests on a causal theory of reference which he takes from Kripke and Putnam. My conclusion *C2* was that it was a mistake for Levine to use a theory of reference to assess *Levine's Thesis*; the direction of assessment should be from biological facts to the theory of reference (5.4). This criticism applied also to LaPorte's and Witteveen's arguments for *Levine's Thesis* and to Brzozowski's argument against.

Still we are interested in semantics as well as biology and so need a theory of reference compatible with the biological facts. So, we need a theory that does not imply *Levine's Thesis*. I argued against the received view that the causal theory does imply this: that's my conclusion *C3* (5.5). A causal theory that includes multiple groundings can explain reference change and accommodate the falsity of *Levine's Thesis*.

The final mistake is about the relation between linguistic decisions and the world (5.6). Haber rightly rejects *Levine's Thesis*, but he does so for the wrong reason. In response to Barry and Jennings' (1998) petition about the MNHN 846 discovery, ICZN (2000) decided to conserve the subspecific names of both *T. S. infernalis* and *T. s. tetrataenia*. Haber thinks that it was this decision that made it the case that 846 had been misidentified as an *infernalis*, hence establishing the falsity of *Levine's Thesis*. Witteveen, who accepts *Levine's Thesis*, has a different view of what that decision achieved: it caused 846 to stop belonging to *infernalis*. It followed from my conclusion *C4* that both these views are wrong: it is a mistake to make any inferences about species identity, and hence about *Levine's Thesis*, from decisions about nomenclature; changing languages does not change worlds. Whether or not 846 is an *infernalis* or a *tetrataenia* and hence has been misidentified is a biological fact that does not depend in any way on a linguistic decision.

I ended my discussion by responding to some objections taken from a couple of unfavorable reviews (5.7). These reviewers wrongly insist on putting linguistic issues center stage in discussing *Levine's Thesis*, despite my argument that this is a mistake (*C2*).

Levine's Thesis is false. So, there would be no paradox even if *Essential Membership* were true. But it is not true (Chapter 4). This does not yield a new paradox. According to *Essential Membership*, MNHN 846 is necessarily a member of its species, *T. s. tetrataenia*. That is quite consistent with the falsity of *Levine's Thesis*: it is consistent with 846 not necessarily being a member of *T. s. infernalis*, the species for which it is a type specimen; indeed, with it not being a member of that species at all.

6
Racial Realism and Essentialism

> Recent biology has confirmed the conviction of those who have long insisted that racial kinds were social kinds, and undermined any possible argument for placing these kinds in the realm of the biological. In its broadest and most common understanding, the concept of race remains little more than the reified residue of racism. (Dupré 2008: 71)

> ...it is hard nowadays to find an unyielding defense of biological racial realism in philosophy. (Maglo 2011: 379)

6.1 Introduction

A major concern in the philosophy of race is whether "race is real", whether "race exists". A related concern is with what races are or, as I would put it, with their essences or natures. My aim in this chapter is to consider these issues from the perspective developed in this book, and in an article, "Natural Kinds and Biological Realisms" (2011c).[1] In contrast to the ringing conclusion from John Dupré quoted above, I shall argue that there are racial kinds, in some sense, that are indeed "in the realm of the biological". Many species have such kinds, including *Homo sapiens*. These racial kinds, like those thought to be part of the Linnaean hierarchy, have essences that are partly historical and partly intrinsic underlying properties.

In claiming this I am emphatically not denying, what Dupré and many others think, that there are racial kinds, in some other sense, that are not proper biological kinds but perhaps social kinds. Dupré offers this example:

> in the UK people are classified as black if they are non-white and hence experience discrimination. It includes people with Asian origins as well as those of African and Afro-Caribbean descent. It would also include (no doubt a very small number of) native Australians or New Zealanders. (2008: 67)

[1] Also in an overlapping article, "Biological Realisms" (2009a).

Biological Essentialism. Michael Devitt, Oxford University Press. © Michael Devitt 2023.
DOI: 10.1093/oso/9780198840282.003.0006

I suspect that this class of blacks is more of a folk-biological kind gone wrong than a social kind. In any case, my main concern is not with such folk kinds nor with whether they are "races". Still, I will discuss them briefly (6.6).

I think the issue of "racial realism" is unclear. The main novelty of this chapter is its attempt to clarify the issue by emphasizing the taxon/category distinction, a distinction that has loomed so large in this book. Another novelty continues a theme from Chapter 5: resistance to bringing claims about language to bear on biological issues. I think that racial realism raises issues that are biologically important and interesting, but an issue that is neither is whether any biological kind, including a human kind, is properly called a "race".

Simply arguing that race is real is not of course endorsing any theory of races, particularly not a racist one. Still, racists are committed to realism about race and so any support for that realism may, willy-nilly, give them some comfort. This raises a terminological issue. The term that white supremacists use for their realism is 'race realism', whereas the term that realist philosophers of race tend to use for theirs is 'racial realism'. I would prefer 'race realism' for stylistic reasons, and because of the comparison I shall make with the issue of "species realism". Still, given the term's nasty associations, I have decided to use 'racial realism' instead.[2]

6.2 A Presentation of the Racial Realism Issue

Quayshawn Spencer begins a recent article:

> In Joshua Glasgow's influential book *A Theory of Race*, he clearly and succinctly defines "racial realism" as the view that "race is real," where "something is real just in case it exists" (Glasgow, 2009: 5).... racial realism might seem to be trivially true. For example, it is widely accepted in the natural sciences that if you can show that even one member of a kind exists, then that kind exists. For instance, that is the strategy chemists use to identify new elements, and it is the strategy that biologists use to identify new species. Given this assumption, and given that, say, Asians, Blacks, Native Americans, and Whites are paradigms of races in current American English, is not it trivially true that race itself exists? Well, it is not that simple. (2018a: 1)

Why is it not that simple? Well, one matter that has to be addressed is what 'race' means. Is it "as understood among biologists (e.g., 'race' as a synonym for

[2] Thanks to Quayshawn Spencer for information and advice on this delicate matter.

'subspecies' or 'ecotype')" or as in "some ordinary language or dialect"? But, either way, Spencer finds philosophers disagreeing about whether "race is real" (p. 2).

There are issues about race that are indeed not that simple but they are not well presented as being about whether race is "real" or "exists". In this respect, they are similar to issues about the Linnaean hierarchy raised by claims like the following: "The species category does not exist" (Ereshefsky 1998: 113); "To the cladist true believer, there is no such thing as a reptile. 'Reptile' does not name a real group" (Sterelny and Griffiths 1999: 197); "taxa of higher rank than species do not exist in the same sense as do species" (Eldredge and Cracraft 1980: 327). According to Ereshefsky this sort of view (which he rejects, 1991: 381) is part of "The Modern Synthesis". He describes that Synthesis as holding: "Higher taxa . . . are merely artifacts of evolution at the species level. So while species are real and the 'units of evolution,' higher taxa are merely aggregates and 'historical entities'" (2001: 229).

What should we make of these claims about what fails to "exist" and be "real"? They all concern kinds of one sort or another. There is an age-old metaphysical issue about the existence or reality of abstract entities like kinds, properties and sets. But, as I have noted (2011c), this "problem of universals" is surely not a "realism" issue that concerns biologists and philosophers of biology.[3] So what is?

6.3 The Reality/Existence Issues about Race

The first crucial step in clarifying this issue is to pay careful attention, once again, to the distinction between taxa and categories. We have seen how particularly important it is to distinguish theories about a *taxon* like dogs—for example, my *Partly Intrinsic Taxon Essentialism* (2.1)—from species concepts, which are theories of what it is for a taxon like dogs to be in the *category*, or *rank*, of **Species**. It is similarly important to distinguish theories of alleged races like Pacific Islanders from theories of what it is for a kind like Pacific Islanders to be in the category **Race**. The former sort of theory is a theory of a *kind* of organism; the latter, a theory of a *kind of a kind* of organism. Discussions of the realism of 'race' tend to be insufficiently attentive to this important distinction between alleged races and **Race**. Demonstrating the importance of this distinction to the racial realism issue is a key aim of this chapter.

Let us start with taxa and with our earlier snake, *Thamnophis sirtalis* (5.3). In what sense does this species "exist" (setting aside whether it exists as a "universal")? Glasgow's familiar idea is that it exists in the sense that it has lots of members that exist; there are lots of San Francisco Garter Snakes, California Red-Sided Garter Snakes, and so on. As Spencer says, "if you can show that even

[3] See Armstrong (1978) for a clear and detailed account of this issue.

one member of a kind exists, then that kind exists". But then, in that same sense, the higher taxon, *Serpentes*, the taxon of all snakes, must surely exist, because it has lots of members that exist, including all the members of *Thamnophis sirtalis*. Similarly, the even higher taxon, *Reptilia* (reptiles) exists because there are lots of crocodiles, snakes, and lizards. I conclude that there shouldn't be *this* realism issue at the taxon level; this realism shouldn't be what the Modern Synthesis is denying. Similarly, there shouldn't be this realism issue about race at the taxon level: at this level, *of course* "race exists" because there are lots of Pacific Islanders, lots of Amerindians, and so on. Antirealists about race couldn't be denying this.

So what could the taxon issue be? Philosophers have been concerned not only with whether kinds exist, in the above sense of having members, but also with whether these kinds are "appropriately special" rather than somewhat arbitrary. In particular, there has been a concern about whether our scientific posits "carve nature at its joints", about whether there is something in the nature of the world that, in some sense, *determines* our categorization of it. I take this to be a concern about whether the kind of entity posited by a theory plays a causally significant role; that is to say, whether it is partly *because* an entity is of that kind that it has the characteristics and behavior that it has. Theories need to posit such kinds if the theories are to be genuinely explanatory. So, there is an issue whether kinds thought to be Linnaean taxa or races qualify: *Are these kinds genuinely explanatory in biology? This, I take it (2011c: 161), is the real issue misleadingly raised by talk of the "existence" or "reality" of taxa and races.*

My "working assumption" from early in this book (2.2) has been that the taxa thought to fall under the categories/ranks of the Linnaean hierarchy are genuinely explanatory. Thus, being a *Thamnophis sirtalis*, a species, features in historical explanations because being that sort of snake is part of the evolutionary story; and that species features in structural explanations because many of the phenotypic properties of a particular snake, type specimen MNHN 846, are explained by its being a *Thamnophis sirtalis*. And one would have thought that there should similarly be no explanatory issue with the higher taxa, despite what cladists and the Modern Synthesis are thought to believe. Thus, being a *Thamnophis*, which is alleged to be a genus, and being a *Bivalvia* (bivalve mollusk), which is alleged to be a class, feature in historical evolutionary explanations and structural explanations, just as being a *Thamnophis sirtalis* does. And so too, being a *Serpentes* and being a *Reptilia*. All these taxa are "real", in this explanatory sense.

This is not to say, of course, that membership in taxa at various levels is *similarly* explanatory: explanatory significance comes in degrees along at least two dimensions. Thus, on one dimension, being a *Thamnophis sirtalis* clearly explains *more of the phenotypic properties of MNHN 846* than does being a *Thamnophis*, which in turn explains more of those properties than being a *Serpentes*, and so on up the hierarchy. On the other dimension, being a *Reptilia* explains *some phenotypic properties of more organisms* than does being a

Serpentes, and so on down the hierarchy. And I take the consensus to be, particularly in evolutionary biology, that being in what is thought to be a species taxon is somehow "at the top" of the explanatory ladder.

The fact that explanatory significance comes in degrees brings home just how misleading it is to raise the issues of the causal roles of taxa and races, hence of the explanatory significance of alluding to them in our theories, by talking of their "reality" or "existence". For, as I have emphasized elsewhere, "*existence does not come in degrees*" (2011c: 159). Snakes exist, hence, in our specified sense, the kind *snake* exists; whereas unicorns do not exist, hence, in that sense, the kind *unicorn* does not exist. There are no degrees about it (though there may occasionally be some indeterminacy).[4] In contrast, kinds can differ enormously in their explanatory significance. Consider cousins, for example (p. 161). Being a *first* cousin, even a *second* cousin, is quite explanatory in biology, leading to interesting generalizations about interbreeding.[5] But what about cousins *in general*? These include not just first and second cousins, but distant relatives like fifth cousins three times removed. The explanatory significance of simply being a cousin of *x* is surely close to zero. And we get even closer to zero if we consider being a step-cousin of *x*. Yet despite this explanatory insignificance, cousins and step-cousins certainly "exist" and are, in that sense, "real".

Despite these objections to using the term 'real' to identify the issue that concerns us, I shall go along with the custom in calling that issue "realism". So, from the perspective of this book, the realism issue about race *at the taxon level* is whether alleged races are like alleged Linnaean taxa in being appropriately explanatory in biology. Clearly, if there are such races, they are kinds below the level of species: they are infraspecific kinds. So we face the question: Are there infraspecific kinds that are thought to be races, even if wrongly thought, and that are appropriately explanatory? In section 6.4, I shall argue that there are in general:

> *Racial Taxon Realism*: There are infraspecific kinds, thought to be races, that are biologically explanatory.

In section 6.5, I shall argue that the human species in particular has such infraspecific kinds:

> *Racial Taxon Realism (Humans)*: There are infraspecific kinds of humans, thought to be races, that are biologically explanatory.

[4] On the indeterminacy, see Devitt (2009b: 57–8).

[5] Kitcher (2003: 245) draws attention to Patrick Bateson's (1982) delightful discovery that the preferred mates of Japanese quail are second cousins.

What about the realism issue about race *at the category level*? This is the issue of whether some of these infraspecific kinds are *rightly* thought to be races. From our perspective, this is also an explanatory issue:

> *Racial Category Realism*: There are infraspecific kinds that exemplify a biologically explanatory category **Race**.

The particular concern will be with humans:

> *Racial Category Realism (Humans)*: There are infraspecific kinds of humans that exemplify a biologically explanatory category **Race**.

Such doctrines have raised the rather overwrought issue: Do these kinds exemplify a category (rank) "worthy of the name" 'race'? I will discuss this in section 6.6. But the doctrine explicitly raises the more theoretically interesting issue: Is the category **Race** explanatory? I will discuss this in section 6.7. I shall argue for *Racial Category Realism*, but not for a "robust" version.

Finally, I have not framed the realism issue, as many do, in terms of whether race is a "natural" kind. Spencer raises the question, "what is actually meant by 'biological racial realism'", and continues:

> So far, philosophers have either meant "race is a natural kind in biology" or "race is a real biological kind", but there are multiple, incompatible views about what a natural kind or a real biological kind is supposed to be. The goal of this article is to defend a new and more promising interpretation of 'biological racial realism' for use in the race debate. That interpretation is that "race is a genuine kind in biology". (2012: 181–2)

The philosophical question of the nature of "natural" (or "real", or "genuine") kinds is a distraction from the racial realism issue. Any acceptable answer to that vexed philosophical question must surely place the explanatory kinds of biology among the "natural" (or "real", or "genuine") ones: the philosophical answer cannot overrule biology. So, the biological racial realism issue should simply focus on whether alleged races are biologically explanatory in the way other taxa are and on whether the **Race** category is biologically explanatory in the way other categories are.

6.4 *Racial Taxon Realism*

Now, we have already implicitly accepted realism for taxa at one infraspecific rank, that of subspecies. Thus, the subspecies *Thamnophis sirtalis tetrataenia* is biologically explanatory (5.3). Whether our explanatory concerns are structural

(1.3) or historical (3.2), it was *because* MNHN 846 was a *T. s. tetrataenia* and not a *T. s. infernalis* that it had its particular "combination of pattern elements" (Boundy and Rossman 1995: 238). The striking phenotypic differences between these two kinds of snake are not sufficient to treat them as distinct species but are nonetheless sufficient to distinguish them by name and to rank them as subspecies. (We shall discuss such rankings in section 6.7.) We should expect, a priori, that there will often be "lesser" differences between groups of organisms within a species, brought about particularly by geographical separation, that warrant distinguishing the groups for explanatory purposes and then assigning them to another rank. And that is exactly what we find. As Alan Templeton notes: "Evolutionary biologists have many words for subdivisions within a species" (2013: 263).

Templeton immediately provides an example: "At the lowest level are demes, local breeding populations" (p. 263). "Varieties" and "forms" provide well-established examples above the level of demes. The *Biology Online Dictionary* describes "Variety" as "a taxonomic rank below subspecies in botany". Similarly, the *Wikipedia* entry for "Variety (botany)" is: "[i]n botanical nomenclature, **variety**...is a taxonomic rank below that of species and subspecies but above that of form". The *International Code of Nomenclature for algae, fungi, and plants* (ICN) (Turland et al. 2018) also recognizes the lower ranks of **subvarieties** and **subforms.**

"Ecotype" is another term applied to groups below subspecies. Its *Wikipedia* entry tells us that it is "closely related" to the just-mentioned "form". It is applied where "phenotypic differences are too few or too subtle to warrant being classified as a subspecies". "*Forma specialis*" is another such term. According to *Wikipedia*, it identifies an "informal grouping" allowed by ICN where authors "do not feel that a subspecies or variety name is appropriate".

The message we should take from this is that biology accepts a hierarchy of infraspecific ranks and uses, informally and a bit loosely, quite a few names for these ranks, from "subspecies" down to "deme" or "subform". The hierarchy is ordered by levels of difference. Biologists take "subspecies" to have the greatest level of difference among infraspecific kinds, "varieties" a lesser level, and so on down to "demes" or "subforms" that have the least.

Are any of the infraspecific ranks thought to be **Race**? Certainly so according to Phillip Kitcher in an influential book:

> under many different names, the idea of intraspecific divisions lingers in ecological and evolutionary studies, where biologists recognize stocks, strains, breeds, evolutionarily stable units, geographical races, morphs, and so forth.
>
> (2003: 255, n. 21)

And certainly so according to Massimo Pigliucci and Jonathan Kaplan in their influential article:

> an awful lot of papers in the nonhuman biological literature have been published in the last five years which include the term 'race' in their title or abstract. (2003: 1162)

They go on to quote the following definition of race from *A Dictionary of Genetics*:

> A phenotypically and/or geographically distinctive subspecific group, composed of individuals inhabiting a defined geographical and/or ecological region, and possessing characteristic phenotypic and gene frequencies that distinguish it from other such groups. The number of racial groups that one wishes to recognize within a species is usually arbitrary but suitable for the purposes under investigation. (King and Stansfield 1990)

Indeed, biologists often apply 'race', again informally and a bit loosely, to infra-specific groups. Thus, the *Wikipedia* entry for "Ecotype" opens with the claim that the term "describes a genetically distinct geographic variety, population *or race* within a species, which is genotypically adapted to specific environmental conditions" (emphasis added). And the entry for "*Forma specialis*" has this to say: "An alternative term in contexts not related to biological nomenclature is **physiological race** (sometimes also given as **biological race**, and in that context treated as synonymous with **biological form**)." The *Dictionary of Botany* relates 'race' to both 'form' and 'ecotype': 'race' "is occasionally used in floras in place of, or subordinate to, form" and "is also used in lieu of ecotype". Finally, consider these excerpts from various entries on **Race** itself. First, from the *Wikipedia* entry on "Race (biology)":

> In biological taxonomy, **race** is an informal rank in the taxonomic hierarchy, below the level of subspecies. It has been used as a higher rank than strain, with several strains making up one race. Various definitions exist. Races may be genetically distinct populations of individuals within the same species, or they may be defined in other ways, e.g. geographically, or physiologically.... The term is recognized by some, but not governed by any of the formal codes of biological nomenclature.... A physiological race may be an ecotype.

The entry pictures the four "races" of the moss *Physcomitrella patens* as examples. Then consider these entries in *Chambers Dictionary of Science and Technology* (Lackie 2007):

> **Biological race**: A race occurring within a taxonomic species, distinguished from the rest of the species by slight or no morphological differences, but by evident differences of habitat, food preference or occupation which inhibit interbreeding.
>
> **Race**: A population, within a species, that is genetically distinct in some way, often geographically separate.

The idea that biological races are ecotypes has had some popularity among philosophers of race (e.g. Pigliucci and Kaplan 2003). So too has the idea that races are subspecies (e.g. Dobzhansky 1944).[6] We have already seen support for the former idea in biology but the latter idea is quite at odds with many of our quotes which place races at a lower level than subspecies. Yet Ernst Mayr (2002) thinks that a subspecies can be a "geographic race" in certain circumstances. Alan Templeton gives the subspecies idea firmer support: "Because of this well-established usage in the evolutionary literature, 'race' and 'subspecies' will be regarded as synonyms from a biological perspective" (2013: 263). And the *Biology Online Dictionary* claims that the only "widely accepted" use of the term 'race' is for "subspecies arising from a partially isolated reproductive population".

In sum, it is very clear that phenotypic, genotypic, and historical differences between interbreeding groups of members of a species have led biologists to posit many infraspecific taxa at a few levels in an infraspecific hierarchy. And 'race' is among the names that biologists use, informally and a bit loosely, for some of these groups. We can abstract one way, though clearly not the only way, that biologists place **Race** in the hierarchy as follows:

1. **Subspecies**
2. **Variety** (particularly in botany)
3. **Form, Ecotype, *Forma Specialis*, Race, Subvariety**
4. **Deme, Subform, Strain**

These infraspecific taxa serve both structural and historical explanatory purposes. *Racial Taxon Realism* is clearly true.

This is not to say, of course, that the *categories* that these taxa are placed in, categories like **Race**, are explanatory; it is not to say that *Racial Category Realism* is true. We shall get to that issue in sections 6.6–6.7. But, given the looseness, we can expect that the category will be a bit vague, at best.

So, what are the natures of these explanatory infraspecific taxa? In my view, they have part intrinsic, part historical, natures like taxa thought to be in the Linnaean hierarchy. The general essentialist doctrine that concluded Chapter 3 applies to taxa alleged to be races:

Taxon Essentialism: The essence of a biological taxon is its members' property of having a certain intrinsic underlying property as a result of descent from members of a certain different taxon, where that taxon has a certain different intrinsic underlying component to its essence.

[6] Adam Hochman (2013) takes this to be the standard view of "racial naturalism", which is my biological racial realism.

My argument for this essentialism about infraspecific taxa is just the same as the earlier explanatory argument for essentialism about Linnaean taxa (Chapters 1–3).[7] So, I will not repeat it, but I shall say more about it in a moment (6.5).

Kitcher rightly makes explicit

> something implicit in the biological practice [of identifying populations as subspecies, or races]: the populations are identified, of course, by phenotypic traits, differences that are sometimes slight, and it is assumed that these differences have arisen over generations of inbreeding. (2007: 296)

This is, in effect, the implicit assumption in biology that there is a *historical* component to the race's essence. I have claimed that

> when biologists group organisms together under some name on the basis of observed similarities, they do so partly *on the assumption that those similarities are to be explained by some intrinsic underlying nature of the group*. It seems to me clear that this is their practice, whatever they say about essentialism.
> (2008; 252–3; see also 1.3 and 2.4.1 above)

This is the implicit assumption in biology that there is an *intrinsic* component to the race's essence.

It is important to see that, on the view presented here, the biologists who posit these infraspecific taxa are not making the simple mistake, mentioned by Kitcher, of rushing in and dividing "organisms according to the differences that strike them as salient" (2003: 242). When all goes well, the biologists are hypothesizing, implicitly and cautiously, that these taxa have identifying intrinsic underlying natures and evolutionary histories that cause these striking differences (1.3).

This concludes my argument for *Racial Taxon Realism*. There are many infraspecific kinds, some known in biology as "races", that are biologically explanatory. These kinds have underlying natures. But what about humans in particular? Are there infraspecific kinds of humans, thought to be races, that are biologically explanatory? Is *Racial Taxon Realism (Humans)* true?

6.5 *Racial Taxon Realism (Humans)*

6.5.1 The Case for *Racial Taxon Realism (Humans)*

The realism question for the human species in particular has, of course, dominated discussions of race outside biology. We have seen that differences among

[7] And Mallon's argument against this race essentialism (forthcoming) is along the same lines as the consensus arguments against that taxon essentialism discussed earlier.

members of a species have often been noted and have motivated distinctions among infraspecific kinds. *Just the same sorts of differences* motivate distinctions that have often been made among humans. The widely used "tree" proposed by Luigi Cavalli-Sforza, Paolo Menozzi, and Alberto Piazza (1994), illustrating the out-of-Africa evolutionary story of humans, seems to exemplify such distinctions. As Spencer sums up:

> if Cavalli-Sforza et al.'s tree is accurate, the human cladistic races that once existed were the following: New Guineans and Australians, Pacific Islanders (excluding New Guineans), Southeast Asians, Northeast Asians (excluding Arctic Northeast Asians), Arctic Northeast Asians, Amerindians, Europeans, non-European Caucasoids, Africans (excluding North Africans)... (2018a: 5)[8]

This is a promising list of human races.[9] These alleged races differ phenotypically in well-known ways and have different evolutionary histories, reflected in their differing geographical origins. It is plausible that their phenotypic differences are caused by genotypic differences, along with the environment of course, and are significant in biological explanation. Still, the racial realist should not, of course, insist on this list of races. With humans, as with any species, it is an empirical biological question which infraspecific kinds are races. And, with humans, some kinds that are very *un*promising candidates to be biologically significant seem to have been thought to be races; for example, the earlier-mentioned (6.1) people "classified as black" in the UK, according to Dupré (2008: 67).

In thinking about the list of races based on Cavalli-Sforza's tree, we should immediately note Templeton's important criticism of such trees. They present a "candelabra" model of human evolution: "the major Old World geographical groups (Europeans, sub-Saharan Africans, and Asians) split from one another and since have had nearly independent evolutionary histories". This stands in contrast to a "trellis" model which posits "recurrent genetic interchange" among populations: "there was no separation of humanity into evolutionary lineages" (Templeton 1999: 636). Yet the evidence favors the trellis model: "The failure of human genetic distances to fit treeness is ubiquitous whenever tested" (p. 639); "The widespread representation of human 'races' as branches on an intraspecific population tree is genetically indefensible and biologically misleading" (p. 646); "our evolutionary history has been dominated by gene flow and admixture that unifies humanity into a single evolutionary lineage" (2013: 700). Nonetheless, Spencer surely has it right in claiming:

[8] Spencer's "once existed" is important. As Andreasen points out, races "are on their way out. With the advent of the modern world came the intermixing of previously isolated populations and the gradual dissolution of racial distinctness" (1998: 200).

[9] Cf. Hardimon's "archetypal examples" of races (2017: 30).

> Cavalli-Sforza et al.'s tree is an idealization that is strictly speaking inaccurate, but accurate enough to be useful in detecting major branchings in the evolutionary history of human populations. (2018a: 7)

Templeton's point is a sign of what has become increasingly obvious in recent years: the differences between alleged human races are not *nearly* as explanatorily significant as the sad history of racism claimed.[10] It was thought (Davenport 1930), on no good basis, that intrinsic underlying differences explained alleged differences in properties central to our view of ourselves as humans: intelligence (IQ), character, and so on. Nonetheless, as Michael Hardimon emphasizes, explaining the familiar and obvious differences in skin color, hair, and facial features between the alleged races is "intrinsically interesting" to biologists (2017: 67).

It is common to think that medical science shows that alleged races play a role in explaining some phenotypic differences that are more interesting than those superficial ones. As Andrew Karter puts it: "There exists adequate evidence of racial disparities to warrant stratification for genomic studies of many complex diseases" (2003: 26). Michael Root, who is critical of the role of biological race in medicine, sums up that role as follows:

> Race is a prominent category in medicine. Epidemiologists describe how rates of morbidity and mortality vary with race, and doctors consider the race of their patients when deciding whether to test them for sickle-cell anemia or what drug to use to treat their hypertension. (2003: 1173)
>
> Most health statistics in the U.S. are stratified by race.... Epidemiologists... typically find that race is a good indicator of the risks of death and disease in the United States.... The rates of many diseases, including major infectious diseases, many cancers, diabetes, asthma and strokes are different between the races.... Heart failure is more common among blacks than whites and symptoms of heart disease develop at an earlier age and progress more rapidly among blacks.... Blacks in the U.S. are seven times more likely to die of tuberculosis than whites and three times more likely to die of HIV/AIDS.... In short, with a few exceptions, people classified as "black" have a poorer health profile than those

[10] And, as Sally Haslanger brings out nicely, racial differences, unlike gender ones, do not provide a basis for different treatments in society:

> Although gender as we know it is a site of social injustice, just societies should be concerned with those functions of human bodies that matter for reproduction. But "color"—those clusters of features such as skin tone, hair texture, eye and lip shape, imagined "racial gene," and other imagined anatomical differences that are used to mark races—does not seem to correlate with any feature that carries sufficient biological weight that it must be socially addressed. (2012: 255)

> classified as "white." So, race, in the view of most epidemiologists, is an important category... (pp. 1173–4)

Root is critical of race's role in medicine for several reasons. He rightly emphasizes that the explanation of medical differences can come from social factors:

> Race is like marital status; no one would be married or single had we not invented matrimony... So too with race; we assign each other a race and treat each other differently depending on that race. As a result, epidemiologists can discover that the rates of mortality or morbidity are different for one race than another. (p. 1175)

But some medical differences between races do not come from how people treat each other, as Root would surely agree; see below on sickle-cell anemia. Such differences are explained by biological differences between alleged races. Furthermore, even where medical differences are primarily explained by discriminatory behavior—heart disease differences are a good example (Wyatt et al. 2003)—biological differences still play *some* explanatory role: the discriminatory behavior is a response, however inappropriate, to biological differences. It is partly because of biologically-rooted phenotypic differences between, say, Amerindians and Europeans, that Amerindians were treated badly.

Sickle-cell anemia is a favorite example of the role of race in medicine.[11] As Root points out:

> Sickle-cell anemia is a recessive genetic disease and is much more common in blacks than whites or Asians in the United States; so common among blacks and rare among whites that for many years sickle-cell anemia has been called a black disease. (2003: 1176)

But, Root notes,

> whites with origins in the malarial regions of Europe are more likely to carry the gene than blacks from regions of Africa in which the risk of malaria is slight. (pp. 1176–7)

[11] Other examples noted by philosophers of race include: cystic fibrosis differences (Dupré 2008); Tay-Sachs in Ashkenazi Jews (Hacking 2005; Kitcher 2007); leukemia (Hacking 2005); lactase persistence differences between Mediterranean, Alpine, and Nordic European "races" (Maglo 2010); aneuploidal fetuses in Asian mothers (Spencer 2019).

But this does not, of course, show that being a black African is not medically explanatory, just that it is not *as* explanatory as being in a subgroup of Africans from a malarial region (or, for that matter, as being in a subgroup of Europeans from a malarial region); explanations come in degrees (6.3). Being a black African yields a biologically-based probability of sickle cell anemia.

That explanations come in degrees is important in responding to another of Root's criticisms:

> Genetic differences do exist between human populations in drug response, but the existing racial categories do not capture these differences very well. Ancestry captures the differences better, since there is less genetic variation within groups identified by common genetic ancestry than groups identified by race. (p. 1179)

Similarly, there is less genetic variation within the species *Thamnophis sirtalis* than within the genus *Thamnophis*. So, as I remarked, "being a *Thamnophis sirtalis* clearly explains *more of the phenotypic properties of MNHN 846* than does being a *Thamnophis*" (6.3). Yet, being a *Thamnophis* is explanatory. Similarly, being an alleged biological race.

Criticisms like Root's are common. Richard Cooper suggests that "groupings of the size of Corsicans, Scandinavians or Yorubans" are more useful than races (2003: 24). And Mike Bamshad claims that "geographic ancestry and explicit genetic information are alternatives to race that appear to be more accurate predictors of genetic risk factors that influence health" (2005: 937). But races can be explanatory and predictive even where smaller groupings, and other information, are more so. And races often are explanatory; see the example of sickle-cell anemia.

In the bad old days, races were thought to be explanatory of a host of very significant properties. That turned out to be quite false. Still, races are explanatory of several superficial properties and a few quite significant ones, like susceptibility to sickle-cell anemia.

One might, of course, accept "realism" about alleged race taxa *in general—Racial Taxon Realism*—whilst resisting "realism" about human races in particular: some other species, like the moss *Physcomitrella patens*, have races but *Homo sapiens* does not; *Racial Taxon Realism (Humans)* is false. This seems a hard position to maintain. The above story about humans is very similar to ones told for many other species. The differences between alleged human races seem just the sort of fairly minor differences between members of a species that have, as noted, prompted biologists to posit the sorts of infraspecific taxa that biologists often call "races": the differences are "few" and "subtle"; the populations are "genetically distinct", "geographically distinct", and so on. Biology should treat humans just like other species (Sesardic 2010). What then could be the basis for resisting racial realism for humans?

6.5.2 Objections to *Racial Taxon Realism (Humans)*

(I) It can be argued that with human the differences are *too* few, *too* subtle, etc.: they are *not significant enough*, to warrant positing different races. Thus, as noted (6.4), it has been common to identify races with subspecies and yet, as Templeton emphasizes, alleged human races do not differ from each other to the extent usually required of subspecies:

> A standard criterion for a subspecies or race in the nonhuman literature under the traditional definition of a subspecies as a geographically circumscribed, sharply differentiated population is to have F_{st} values of at least 0.25 to 0.30 (Smith et al. 1997). Hence, as judged by the criterion in the nonhuman literature, the human F_{st} value is too small to have taxonomic significance under the traditional subspecies definition. (1999: 633)

Indeed, the F_{st} value for the five major "races" of humans was discovered to be only 0.043. This stands in striking contrast to a value of 0.301 for chimpanzees (Templeton 2013: 267).

The fact that these alleged races do not differ enough to meet this (apparently arbitrary) 0.25 criterion for subspecies does not show, of course, that they do not differ at all. Similarly, as Templeton notes, the apparent earlier-noted fact that humans form a single evolutionary lineage

> does not mean that all human populations are genetically identical. Past founder events, isolation-by-distance, and other restrictions on gene flow ensure that human populations are genetically differentiated from one another, and local adaptation ensures that some of these differences reflect adaptive evolution to the environmental heterogeneity that our globally distributed species experiences. (2013: 270)

Still, Templeton continues, "most of our genetic variation exists as differences among individuals, with between population differences being very small (p. 270). As Kitcher notes succinctly, it is common to argue against human races (e.g. Maglo 2011) by claiming that "intraracial diversity is far more pronounced than interracial diversity" (2003: 230); the *locus classicus* is Lewontin (1972). This argument is notably reminiscent of Leslie's claim, discussed in subsection 2.4.2: "a member of one species... may have more genetically in common with a member of another species... than with a member of its own species" (2013: 133). And my response to the race claim is similar to my earlier one to the species claim. In brief, what matters to positing a race is not overall similarities among members and differences from non-members, but a similarity in one particular part of the underlying structure, the part that causes the distinctive phenotypic features of the race. Pigliucci and Kaplan make a related point:

> The question is not whether there are significant levels of between-population genetic variation overall, but whether there is variation in genes associated with significant adaptive differences between populations... (2003: 1165)

And there surely is. Thus, Kaplan introduces a very helpful discussion of 'Race' in the *Encyclopedia of Life Sciences* as follows:

> That human populations differ in the frequencies of particular alleles, and that these differences are not uniform or random but follow patterns associated with the ease and historical frequency of gene flow (as well as with local selection pressures), is not in doubt. (2011: 1)

And he concludes his discussion:

> That (some) populations [that we call 'races'] do differ in allele frequencies, and hence are 'biological races' as well as social races is not much in doubt. (p. 5)

(II) In the context of rejecting the idea that races are ecotypes, Templeton raises another problem for *Racial Taxon Realism (Humans)*:

> All species, humans included, adapt to many environmental factors, not just one. Frequently, different adaptive traits display discordant geographical distributions because the underlying environmental factors have discordant geographical distributions. As a result, one will get different ecotypes for different adaptations.... Depending upon which adaptive trait is chosen, one will get very different "races". So which adaptive traits should be used and which should be ignored? Evolutionary biology provides no objective way of addressing the question of choice of adaptive traits, so the ecotype concept of race in humans is yet another subjective, culturally sensitive concept of "race". (2013: 269)

The problem here is not (I)'s problem that groups alleged to be races differ in only minor ways but that it is biologically *arbitrary* to pick out those groups rather than others that also differ in minor ways.

I start my response to this arbitrariness concern by emphasizing three points about kinds, drawing on the discussion in section 6.3. First, any species can be divided into countless infraspecific kinds each of which will exist in our minimal sense of having members. Second, among those countless kinds will be some that are explanatory in that being a member of one explains the distinctive phenotypic properties of members of that one. Third, explanation comes in degrees and so some of these kinds are more explanatory than others (on one or another dimension). Douglas Futuyma notes that

> differences among populations range from the immeasurably small, through varying degrees of differentiation, to levels of behavioral, chromosomal, and developmental distinctiveness that are characteristic of different species.
>
> (1986: 103)

Now any kind at *any* infraspecific level is worth adverting to in our theory if it has *sufficient* explanatory power. We posit an infraspecific kind when we suppose that its members share an underlying property and an evolutionary history that explains significant shared traits that distinguish it from other members of its species. In this, *no* adaptive trait should be ignored. There is no reason to be miserly here: if a kind is sufficiently explanatory, it should be given a place in the theory, even if only a small place. The evidence is that biologists are not miserly: where they have an explanatory need, they posit kinds, at various infraspecific levels, as we have seen (6.4). And they call some of them "races". Alleged races are sufficiently explanatory kinds at a somewhat vaguely identified infraspecific level. They are not arbitrary.

This positing of kinds to explain differences, even down to quite small differences, is of course common in science. Consider linguistics, for example. We posit a linguistic "family", Slavic, that includes many "languages" that have a great deal in common. One of those languages is Serbo-Croatian. But the languages spoken by Serbs and Croats differ a bit. Much has been made of this difference in recent years, for political reasons. Still, politics aside, there is an explanatory basis for distinguishing Serbian from Croatian, even if not for labeling them different "languages" (Greenberg 2004). Differences within Croatian led to the positing of the Chakavian "dialect" spoken all along the coast of Croatia (Lisac 2009). And the need to explain differences did not stop there. "Local dialects" or "vernaculars" of Chakavian have been identified and studied by linguists; for example, one spoken in the city of Split (Jutronić 2010); and even one in the small village of Sutivan (Jutronić and Galović, forthcoming).[12]

(III) Another concern about *Racial Taxon Realism (Humans)* is that "human variation is clinally distributed.... Allele frequencies change gradually across geographic space, with few sharp discontinuities" (Bolnik 2008: 72). "This makes the choice of where one 'race' ends and another begins largely arbitrary" (Hochman 2021: 454). This is reminiscent of concerns about species discussed in sections 1.10 and 2.4.7.

[12] My thanks to Dunja Jutronić for this information.

First, there is a concern about *indeterminacy* (or vagueness). It is somewhat indeterminate where to draw the line between an alleged species and its ancestor. More to the point, with ring species it is somewhat indeterminate where to draw the line between an alleged species and its neighbor. Similarly, it is somewhat indeterminate where to draw the line between an alleged race and its neighbor. But "indeterminacy is everywhere" (2008: 373; 1.10 above). It is indeterminate where to draw the line between mountains and non-mountains, but there really are mountains and it is explanatory to be one. Similarly, despite indeterminacy, there really are Amerindians and it is explanatory to be one.

Second, there is a concern about *arbitrariness*. There may indeed be some arbitrariness with alleged races: it may be somewhat arbitrary which somewhat indeterminate group to pick out as a race just as it may be which somewhat indeterminate group to pick out as a species (2008: 374; 1.10 above). But the arbitrariness should not be exaggerated: our decision about which group to pick out here, as elsewhere in science, is driven by our explanatory purposes. Just as it is explanatory for an organism to be a member of a somewhat indeterminate group, picked out perhaps a little bit arbitrarily, as a species, so too is it, as a race. These groups are real, despite indeterminacy and some possible arbitrariness in picking them out for our scientific purposes.

Two points need emphasis. First, the distinction between indeterminacy and arbitrariness is important. Kinds *will be* a bit indeterminate in that there is no fact of the matter whether or not certain entities are members of them; for example, as we have often noted (quoting Hull 1978: 306), the taxon *horse* is a bit indeterminate because there is no fact of the matter whether or not certain organisms ancestral to current horses are horses. Kinds *may be* arbitrary in that there is no persuasive explanatory reason for a theory naming one (indeterminate) kind rather than another that overlaps with it; think of anagenesis with gradual evolution (2008: 374; 1.10 above). Second, we are concerned here with *Racial Taxon Realism (Humans)*, with whether there are explanatory subgroups of humans that are thought to be races. We are not concerned with *Racial Category Realism (Humans)*, with whether they are *rightly* thought to be races. That category issue generates concerns about indeterminacy and arbitrariness too, as we shall see in subsection 6.7.4.

(IV) Finally, consider Dupré's ringing conclusion that heads this chapter:

> Recent biology has confirmed the conviction of those who have long insisted that racial kinds were social kinds, and undermined any possible argument for placing these kinds in the realm of the biological. (Dupré 2008: 71)

He is surely very well aware of the talk of race in biology illustrated in section 6.4. So what explains his opinion? Dupré supposes that racial realism requires that there be a "genetic basis for race", "a distinctive genetic inheritance to traditional racial groups" (2008: 50). Dupré produces a lengthy, and wonderfully informative, exploration of the underlying causes of phenotypic properties, concluding that "there are no 'genes for' race in any of the various senses of the word 'gene'" (2008: 70). Similarly, Kitcher claims, "there are no genes distinctive of the groups we call races" (2007: 294). Dupré and Kitcher think that there is no genetic basis for membership in any of the kinds alleged to be races. So, it seems, they must think that though biologists may occasionally informally call some kinds "races", those kinds are not really races.

My essentialism about races is, of course, committed to (near enough) what Dupré and Kitcher are denying. For, races are like the Linnaean taxa. They have their causal-explanatory roles because of their essences. I have argued for this view of their essences:

> *Taxon Essentialism*: The essence of a biological taxon is its members' property of having a certain intrinsic underlying property as a result of descent from members of a certain different taxon, where that taxon has a certain different intrinsic underlying component to its essence. (3.7)

Taxa have their place in the causal nexus in virtue of these essences, thus meeting what we called "the Sober demand" (3.2). The essence of a taxon, along with the environment, cause the phenotypic properties of the taxon.[13]

Dupré's discussion does not count against the racial taxon essentialism that that I am urging. As I emphasized in responding to Leslie (2.4.5), this sort of essentialism is committed to intrinsic underlying essential properties that are "probably largely, although not entirely genetic" (2008: 347; 1.1 above), with the details to be discovered by biologists. Dupré's discussion, with its emphasis on the complexity

[13] It is interested to note that this is a modern version of a view of the "racial essence" that Anthony Appiah claims was held by "most Western scientists (indeed, most educated Westerners)" "by the end of the nineteenth century":

> to say someone is "Negro" is not just to say that they have inherited a black skin or curly hair: it is to say that their skin color goes along with other important inherited characteristics. (1996: 466)

It is also interesting to note what, according to Faucher, was thought to be the essentialist view held by the folk:

> Until recently, researchers mostly attributed the following conception of race to the folk... Human races are social groups in which
> 1. Individuals share a number of physical and psychological features that are specific to their group and that they do not share with any other group;
> 2. The fact that they exhibit these features is explained by the presence of an underlying and unobservable cause, an "essence";
> 3. The possession of this essence is necessary and sufficient for membership in the group; and
> 4. They share these features in virtue of a biological mechanism that ensures the transmission of the racial essence from generation to generation. (2017: 249–50)

of underlying causes (2008: 53–6), gives further support to the earlier point that "not only is the essence of tigers not 'a tiger gene', it is a long way from a tiger gene" (2.4.6). A reasonable essentialism is not committed to *simple* genetic differences between races any more than between species. Passages like the following are thoroughly congenial to such essentialism:

> it is in general quite mistaken to think of bits of the genome having specific functions defined in terms of phenotypic outcomes. So, finally, there is no reason to expect a particular set of genomic features to provide a complete causal explanation of a feature such as skin color. Like other features, we should expect skin color to be the final outcome of various possible developmental pathways, exploiting a range of genomic and other developmental resources.
>
> (Dupré 2008: 62–3)

The essentialism I am urging claims that there are underlying essential properties "at the genetic level", however complex, that, along with the environment, are causally responsible, perhaps by several developmental pathways (2.4.6), for the phenotypic properties characteristic of groups thought to be races.[14] This is not undermined by Dupré's discussion. Nor does it amount to what Kitcher calls "crude genetic determinism" (Kitcher 2007: 305, n. 22).

Furthermore, just as this sort of essentialism is plausible for Linnaean taxa (2.4.1), so too is it for alleged races. How so? Consider any person with the characteristic phenotypic properties of an alleged race. What explains the fact that she developed into a person with those properties? Some intrinsic underlying property at the genetic level together with the environment. That is surely undeniable. Then the plausible essentialist hypothesis is that the *very same* underlying property, whatever it may be, explains those phenotypic properties in other members of her race in that environment. As noted in section 6.4, I take this hypothesis to be implicit in biological practice.

Indeed, applying an earlier point about Linnaean taxa (2.4.1), what plausible alternative is there to the idea that there is a common underlying cause of the phenotypic properties used to identify a race? Surely not that there are a large variety of underlying causes. We might, of course, discover that more than one distinct underlying nature is causally operative in an alleged race. Thus, though "Melanesians and Africans share dark skin, hair texture, and cranial-facial morphology . . . the traits typically used to classify people into races", "these two human populations have nearly maximal genetic divergence within humanity as a whole with respect to molecular markers" (Templeton 1999: 640). So, lumping Melanesians and Africans together in the same race would be a biological mistake

[14] The interaction between the genotype and the environment in causing phenotypic properties is also complex, as Pigliucci emphasizes (2013).

analogous to that made with cryptic species, a mistake that should lead to a *re*classification (2.4.1). The lumping together of all the people "classified as black" in the UK is, of course, an even bigger biological mistake. All this shows that we can be wrong about what groups are races. It does not show that we are wrong about race essentialism.

6.5.3 Rosenberg et al. and the Essences of Alleged Races

I have made no commitment to a precise kind of intrinsic property constituting the essence of a biological taxon. As I have emphasized, "the wise philosophical essentialist leaves the details of essences to scientists; biological essences are to be discovered by biologists" (2.4.5). Essentialism about race does not require commitment to these details. Still, an influential study by Noah Rosenberg et al. (2002) throws some light on the details.

This study has, in effect, been brought to bear primarily on the category question in the racial realism debate, on *Racial Category Realism (Humans)*. Are there infraspecific human taxa exemplifying an explanatory category (rank) "worthy of the name" 'race'? We shall consider this role for Rosenberg et al. in the next section (6.6). However, those discoveries do bear on our taxon question.

Rosenberg et al. "studied human population structure using genotypes at 377 autosomal microsatellite loci in 1056 individuals from 52 populations" (2002: 2381). Their procedure "places individuals into *K* clusters", where '*K*' is a numeral.

> At $K = 2$ the clusters were anchored by Africa and America, regions separated by a relatively large genetic distance.... Each increase in *K* split one of the clusters obtained with the previous value. At $K = 5$, clusters corresponded largely to major geographic regions. (p. 2382)
>
> Genetic clusters often corresponded closely to predefined regional or population groups or to collections of geographically and linguistically similar populations. (p. 2384)

Thus, as Spencer points out, at $K = 5$ the clusters corresponded to "Africans, Eurasians, East Asians, Oceanians, and Native Americans" (2019: 98). And, he observes,

> we have a causal explanation for how [these continental clusters] arose. First, humans migrated to different major geographic regions (e.g. East Asia, Oceania, the Americas, etc.). Next, major geographic barriers to human interbreeding (e.g. the Himalayas, the Pacific Ocean, etc.) prevented random or even frequent interbreeding among people after these new lands were occupied. Finally,

> genomic diversity arose and spread among people in the same continental regions (e.g. from mutation and interbreeding) after enough time passed (Rosenberg, 2011, 681).[15] (Spencer 2018b: 10–11)

Spencer argues "that the set of human continental populations is biologically real because it currently occupies the K = 5 level of human population structure according to contemporary population genetics" (2019: 103).

I agree with Spencer that we can see Rosenberg et al. as providing *evidence* in support of realism about race taxa. From my perspective, the continental genetic clusters, reflecting genetic similarities, are evidence that each of these taxa have an appropriate essence, an essence that can feature in structural and historical explanations of the characteristic phenotypic properties of the taxon. For, the clusters are evidence that these taxa have what could play this explanatory role: an intrinsic underlying essence at the genetic level with an appropriate evolutionary history. This is good news for the realist but, I emphasize, the discoveries of Rosenberg et al. do not themselves *provide* the needed essence. We have evidence that the intrinsic essence is to be found somewhere in the taxon's genetic cluster but we are not told where. Locating the essence requires locating which aspects of the cluster *play the required causal-explanatory role.* As Kitcher says:

> the clusters demarcated on the basis of genetic similarity are not going to play a significant role in the explanation of shared phenotypic features or susceptibilities to various types of disease. (2007: 305)

So races cannot be *identified* with these continental genetic clusters. Koffi Maglo is right to "deny . . . that continental genetic partitions *constitute* natural subdivisions or, more exactly, evolutionary kinds" (2011: 370; emphasis added).[16] Still, we have evidence here that these continental populations thought to be races are biologically explanatory, and not just explanatory in evolutionary/historical explanations but also in *structural* ones.

I shall turn now to the realism question at the category level, particularly for humans, *Racial Category Realism (Humans).* As noted in section 6.3, there has been a tendency to blur the distinction between the category and taxon question. Furthermore, the focus has been, in effect, on the category question. Here is just

[15] Of course, given the earlier-noted evidence for the "trellis" rather than "candelabra" model of human evolution (6.5.1), these geographic barriers to human interbreeding did not prevent "recurrent evolutionary interchange" (Templeton 2013: 636).

[16] And Hochman is right to say:

> The fact that individuals from geographically distant and not recently admixed populations can be clustered according to their continental origins does not, on its own, constitute an argument for human subspecies division. (2013: 343)

one example,[17] which I take to be typical, from an excellent article by Adam Hochman. He defines "racial naturalism"—what I am calling biological "racial realism"—as "the view that humans can be divided into subspecies, and that 'race' is therefore a valid scientific category" (2013: 342). There are two distinct theses underlying this naturalism/realism. One is the taxon thesis that humans can be divided into infraspecific taxa for scientific purposes, in effect, the thesis *Racial Taxon Realism (Humans)*. The other, which Hochman seems to mostly have in mind, is the category thesis that any such taxa can be put in a valid scientific category of **Subspecies**, in effect, a version of the thesis *Racial Category Realism (Humans)*. And, in my view, the most persuasive arguments against racial realism are not the ones just considered against the taxon thesis but the ones to be considered below against the category thesis. Yet, I shall suggest, the taxon thesis matters more than the category thesis. The plausible division of organisms into infraspecific kinds is biologically important, even if not nearly as important as the division of them into species. For example, it is biologically important that the human species is made up of Pacific Islanders, Amerindians, and so on, kinds that differ in small ways that are explanatorily significant. These kinds pose some interesting category issues but these are not, it seems to me, as important as the taxon thesis.

6.6 *Racial Category Realism (Humans)*: The "Worthy of the Name" Issue

What could the realism issue about human races be at the category level? As already noted, it is surely not the issue, raised by the age-old "problem of universals", of whether the *universal* **Race** exists. I take the issue to be:

> *Racial Category Realism (Humans)*: There are infraspecific kinds of humans that exemplify a biologically explanatory category **Race**.

Now clearly if *Racial Taxon Realism (Humans)* were false, if there were indeed no biologically explanatory infraspecific kinds of humans thought to be races, then *Racial Category Realism (Humans)* would be false too. But once the taxon doctrine has been accepted, we face the category doctrine. This explicitly raises an explanatory issue about the category **Race**, which we will consider in section 6.7. But first we must consider what I described (6.3) as "the rather overwrought issue" of whether these infraspecific human kinds called "races" are "worthy of the name"

[17] Other examples include Andreasen (1998: 201); Spencer (2018a); Mallon (2018: 1042–3).

'race'. *I shall argue for Racial Category Realism (Humans)*, but not for a "robust" version.

We have seen that Dupré's opposition to biological racial kinds rests partly, at least, on his rejection of a genetic basis for races (6.5.2). I think it also rests partly on his conviction that, in its *ordinary* meaning, the term 'race' refers to social kinds (2008: 61, 66–9) and so should not be applied to biological kinds. Indeed, many are concerned that, because of its ordinary meaning, 'race' should not be so applied, as Luc Faucher indicates:

> Biophilosophy then faces a challenge: should the term "race" be retained even though the entities to which it refers do not have the properties attributed to them by folk notions, or should we simply drop the term? (2017: 248)

Joshua Glasgow is a good example of the concern. He is critical of the idea that "the cladistic approach" of Robin Andreasen (1998)

> affords race some biological reality.... it seems that the nine races in [Cavalli-Sforza's tree] do not correspond extensionally with folk notions.... Andreasen takes this discontinuity with the folk notion of 'race' to be nonproblematic... But this exposes what I take to be a central flaw of the cladistic approach. That is, Andreasen has found a way of carving our ancestors into breeding populations, but these populations are not what we call 'races'. (Glasgow 2003: 458–9)

Later: "Andreasen's theory has not shown that race *in its usual meaning* is real" (p. 462; original emphasis). Kitcher's view comes in for similar criticism: "I do not think that this picture is sufficiently faithful to the folk category" (p. 467).

This concern with the ordinary meaning of the term 'race', or the ordinary concept of race, is understandable because the term's and concept's association with racism, and hence historical roles as instruments of discrimination and oppression, have made them rather toxic. Any suggestion that the category identified by that term/concept is biologically real may seem to be giving credence to obnoxious racist theories. But this is a mistake. The reality of the category **Race** is one thing, theories of races are another very different thing; the *concept* of race is not a *conception* of races.[18] Furthermore, this concern with the ordinary meaning

[18] It is also a mistake to suggest that racism is based on racial essentialism:

> racial essentialism... is a *bad* view because the false belief that such essences do exist and play the proposed explanatory roles seems to provide a metaphysical basis for generalizing about members of a race in pernicious ways. (Mallon forthcoming)

Essences, along with the environment, are the metaphysical basis for *all* generalizations about *all* kinds, including races. So, the badness of racist generalizations could not come from believing that races have essences. The badness comes from false theorizing about *which* generalizations are based on the (largely unknown) essences.

of 'race', and the ordinary concept of race, is misguided, as I shall briefly try to show. The little we really know about these matters, as indeed about *any* ordinary meanings or concepts (Devitt 2021b)—we should be wary of the idea that we can do semantics from the armchair[19]—does not motivate the concern.

(A) Let us start by supposing the worst about the meaning of 'race' for the racial realism case I am presenting: that in its ordinary meaning 'race' does not apply to biological kinds but rather to social kinds. This does not show that 'race' does not have another, "more scientific", meaning in which it *does* refer to biological kinds; it might be, in this way, ambiguous, as indeed Andreasen points out (2005: 105) and Spencer endorses (2018a: 5). Dupré (1981) himself was a pioneer in suggesting that such ambiguities are common in the words for living things.[20] He noted that these terms have functions for the folk that differ from those of scientific terms; folk terms are for kinds that are "economically or sociologically important", "furry and empathetic", "very noticeable" (p. 80), and so on. So, Race might be socially real in one sense and biologically real in another.[21]

If the biological meaning of 'race' was *unrelated* to the ordinary one, if the ambiguity was a matter of homonymy not polysemy—like 'bank' meaning river sides and financial institutions—then it might be appropriate to claim that the biological kinds are not "worthy of the name" 'race'. It might be appropriate to say that in arguing for races in this biological sense we are not arguing for *racial realism* but changing the subject. I take it that this is Glasgow's concern:

> the core question: How revisionist can one be about the meaning of 'race' and still call it 'race'? . . . while some minimal revision to the meaning of 'race' . . . is allowable in the search for biological backing for race, we must stay fairly close to the vest, or we risk not talking about race at all . . . the burden is on the revisionist to show that her revisions are warranted. (2003: 462–3)

Yet the use of 'race' in biology clearly does not make 'race' homonymous.[22] Thus, note that the following is among the seven (most relevant) "senses" in the *OED*: "A breed or stock of animals; a particular variety of a species" (first cited usage was in 1580). And the seven are listed together under the following heading: "A group of persons, animals, or plants, connected by common descent or origin". These senses are clearly related to the biological meaning we have demonstrated. So too is the presumably ordinary meaning of 'race' exemplified by the widely used

[19] So, we should be wary of Glasgow's emphasis throughout his paper on "the significant degree to which the intension of 'race' is tied to gross morphological features, such as skin color" (2003: 472).

[20] See Devitt and Porter (2021) for a report on recent empirical investigations of this issue.

[21] Cf. "There are three central positions within this debate: race might be biologically real, socially real, or not real at all" (Hochman 2014: 83).

[22] Though its use to mean a certain sort of competition—for example, "a 100m race"—may make it homonymous.

distinctions between races made by Office of Management and Budget (OMB) largely on the basis of the geographical origin of their "original peoples" (1997). So too is the meaning exemplified in the description of nineteenth-century views given in note 13. So too is the ordinary meaning discussed in detail by Michael Hardimon (2017). We should conclude that if 'race' is really ambiguous, having a biological meaning as well as an ordinary meaning, it is clearly polysemous. And there is then no basis for saying that the kinds biologists are talking about are so unlike what we ordinarily mean by 'race' as to be unworthy of that name. As Sally Haslanger neatly remarks, "there is no Language Police to judge that the biologist is wrong to use the term 'race' in a particular way" (2012: 299).

(B) *Does* 'race' in its ordinary meaning apply to social kinds? Perhaps not. Perhaps it applies to social kinds in some ordinary *uses* but has broader application in its ordinary *meaning*. We should not presume that all uses, like the *OED*'s seven "senses", are to be treated theoretically as distinct *conventional meanings*. It is a tricky theoretical question to determine whether such uses constitute distinct meanings or rather, as Charles Ruhl (1989) and many other linguistic pragmatists/contextualists would likely think, *pragmatic modulations* of the one, more abstract, meaning.[23] Perhaps 'race' is not a case of polysemous ambiguity but rather of a vague, underspecified monosemy: "questions concerning polysemy and monosemy are some of the most fundamental in lexical semantics" (Cruse 1992: 577). So maybe the biological use of 'race', like the social use, is just one among several acceptable modifications of the one abstract ordinary meaning.

(C) That's one reason for caution about thinking that, in its ordinary meaning, the term 'race' applies to social rather than biological kinds. Here are some more. Why suppose in the first place that the ordinary term *does* refer to social not biological kinds? Dupré (2008) puts his finger on what I take to be the main reason. He points out, and I agreed (6.1), that the kind identified as "blacks" in the UK is not a proper biological kind, not an explanatory kind of the sort adverted to in biological theory. This is true of "blacks" in America too. And it is probably true also of the Irish, who were often picked out as a race in nineteenth-century America (and later).[24] But, importantly, the fact that these kinds are not proper biological kinds does not alone show that they are *social* kinds. Perhaps they are *im*proper biological kinds, having their place in *failed* biological explanations of phenotypic properties; perhaps they are biological kinds "gone wrong", as I suggested in section 6.1.

(D) Set that aside and suppose that they are social kinds. That fact would *still* leave us quite a way from establishing that 'race' ordinarily refers to social kinds.

[23] I argue (2021b) that Ruhl's "monosemantic bias" is quite wrong, but this is certainly not something that can be taken for granted.

[24] "The Irish Race in America" was the name of a book (by Edward O'Meagher Condon) published in 1923.

Indeed, how might the allegedly social nature of kinds like the UK blacks and the Irish bear on the ordinary meaning of 'race'? Presumably, because the folk are thought to apply the term 'race' to these kinds:

> What "we" in public discourse call race is not a natural or genetic category. Rather, the ordinary term 'race' picks out a social type, that is, the objective type that attracts our reference is unified by social features rather than natural ones. (Haslanger 2012: 307)

Suppose that we really do call at least some social kinds "races". This would still not show that 'race' refers to social kinds. The application may be a *mis*use of the term, given the meaning of the term in the folk's very own language. "Ignorance and error" arguments (Devitt and Sterelny 1999), first proposed by Kripke (1980) in discussing proper names like 'Columbus', 'Einstein', and 'Cicero, show vividly that a term can be part of a person's language even though she is quite unable to successfully identify the entities it refers to.

A similar response is appropriate to Kaplan's claim that "the populations called 'races' in ordinary social discourse" are social kinds (2011: 1). We noted his view that what the folk call "races" do differ in biologically significant ways (6.5). Nonetheless, he insists, so too do many other groups of humans: "folk-racial categories" are "only one of many ways of dividing up populations into biologically meaningful groups" (p. 5). So, "we cannot uniquely map the biological categories that emerge onto our folk-racial categories" (p. 3). So, the folk categories are not vindicated; they "remain social categories and not biological categories" (p. 1). But there is an alternative view: the folk's term 'race' refers to biological kinds but the folk are not too good at identifying races and so miss a few.

(E) Finally, even if the kinds in question are social, perhaps the folk do not think of them all as *races* and so do not apply the term 'race' to them all. Many social kinds, for example Tories, are not taken to be races. More to the point, think of Hispanics. Many Americans probably regard them (rightly) as forming an *ethnic* not racial group; and most people who self-report as Hispanic do not regard this as their race (Taylor et al. 2012).

In light of points (B) to (E), it is far from obvious that the ordinary meaning of 'race' does refer to social not biological kinds. Perhaps its ordinary meaning is abstract with an acceptable modification to refer to a biological kind as well as one to refer to a social kind. Perhaps some of the kinds that 'race' is applied to by the folk—for example, UK blacks—are not social kinds but rather improper biological kinds. Alternatively, perhaps those kinds are social, but the folk's application of 'race' to them is a misuse of the term in their language. Finally, perhaps the folk do not apply the term to these assumed social kinds. So, the view that the ordinary meaning of 'race' refers to social kinds rests on several questionable semantic assumptions. Still, maybe that view is right. If so, the case that the infraspecific kinds that biologists call "races" are "worthy of the name" rests on (A). Calling those kinds "races" exemplifies a usage of 'race' that is, at least, related to an ordinary one. In sum, the little we know about the meaning of 'race' supports the name-worthiness.

We should wonder about the importance of this "semantic" issue to racial realism. It matters that the human species is divided into explanatorily significant infraspecific kinds. The categories of these infraspecific kinds are also of some interest, as we shall see. But whether a category is properly called "race" surely does not warrant much attention. What's in a name? (The name-worthiness issue returns in section 6.8.)

The ordinary meaning of the word 'race', and the ordinary concept of race which that word expresses, receive a lot of attention in the race debate.[25] If our concern was semantic, this would of course be appropriate. Perhaps also if our concern was social.[26] But given that our concern is biological, it seems to me that the issues raised in (A) to (E) exhaust the interest of this ordinary meaning and concept. The ordinary meaning and concept have their place in the folk theory of race. Insofar as that folk theory is in accord with our best science then we should stick with it and its concepts. Insofar as it is not, we should abandon it and perhaps even its concepts.

In sum, consideration of the ordinary meaning of the term 'race' does not undermine the idea that some infraspecific taxa exemplify a category (rank) "worthy of the name" 'race'.

6.7 *Racial Category Realism (Humans)*: The Explanatory Issue

6.7.1 The Explanatory Issue with Other Categories

So much for the first realism issue of race at the category level. It is helpful to approach the second, and much more important, issue by considering discussions of the reality/existence of other categories.

Start with Ereshefsky's earlier-quoted claim, "The species category does not exist" (1998: 113). What are we to make of this? It is explicitly not a claim about species taxa; it is not about "the reality of those lineages we call 'species'" (p. 104). I have argued that we should take the claim as "denying the *explanatory significance* of kinds being species" (2011c: 161). Ereshefsky's denial stems from his pluralism about species concepts, noted in section 4.8. He thinks that there are

[25] Thus, Spencer has an article entitled "Are Folk Races Real?" (2018b) and Hardimon defends the biological reality of race in "ordinary uses of the English word 'race' and its cognates" (2017: 27).

[26] Haslanger claims that "'race'... plays such a major role in our self-understandings and political life" (2012: 304). Perhaps so, but I do wonder whether the major role is played by words like 'black' for alleged racial kinds rather than by the word 'race' for the category of such kinds.

several explanatory roles played by taxa in virtue of being in a species category and so there are several equally good categories: there are "biospecies", "phylospecies", and "ecospecies" (1998: 117). Although it is explanatorily significant for a taxon to be in one of these categories it is not for it to be a "species", which is a mere disjunction of the significant categories with no "distinctive commonality" (p. 115). As I construe Ereshefsky, this is the respect in which the categories of biospecies, phylospecies, and ecospecies "exist" but the category of species does not.

So Ereshefsky, and some others (including Baum 2009; Dupré 1999; Kitcher 1984; Mishler and Donoghue 1982; Sterelny and Griffiths 1999), have doubts about the category, **Species**. But Ereshefsky, and just about everyone, have doubts about the acceptability and reality of the higher categories/ranks. Ereshefsky rightly points out that if a certain category is to be acceptable, the taxa that fall under that category should be "comparable", and he draws attention to reasons for thinking that this condition is not met (1999: 299). Brent Mishler claims that "practicing systematists know that groups given the same rank across biology are not comparable in any way" (1999: 310–11). In a lengthy critique of the Linnaean hierarchy, Ereshefsky mentions disagreements about the rank of certain taxa (2001: 226); and the drive to introduce more ranks, leading to a hierarchy in flux (p. 215). He cites evidence "that Darwin doubted the existence of the Linnaean categories" (2001: 231). Sterelny and Griffiths (1999) have this to say about the higher categories. "[W]hat is a genus? A family?" On such issues, "[s]ystematics has gone through a long period of controversy, some of it extraordinarily bitter" (p. 194). They describe the various taxonomies, phenetic, evolutionary, and cladistic, pointing out that the evolutionary and the phenetic taxonomies may group lineages into genera differently (p. 196; see also Ereshefsky 2001: 226). They think that cladism "presents the best view of systematics" (p. 200), but from that perspective, the higher ranks "make little sense":

> [cladists] do not think there will be any robust answer to the questions when should we call a monophyletic group of species a genus? a family? an order? Only monophyletic groups should be called anything, for only they are well-defined chunks of the tree. But only silence greets the question are the chimps plus humans a genus? It has long been received wisdom in taxonomy that there is something arbitrary about taxonomic classification above the species. These decisions are judgment calls. So cladists show only a somewhat more extreme version of a skepticism that has long existed. (p. 201)

Ereshefsky (1999, 2001) and Mishler (1999) argue persuasively that the Linnaean hierarchy of categories should be abandoned.

In light of the controversy over the species category, it is helpful to describe this skepticism in terms of "concepts", in terms of theories (definitions) of what it is for a taxon to be in a certain higher category. Thus, the cladists think that there are no "robust" concepts for the higher categories. Biologists generally have not been convinced by any concept that has been proposed. So, there is widespread skepticism in biology about the higher categories.

I emphasize that this issue with the higher categories, as with **Species**, is not about the explanatory significance of the taxa that are thought to be in a category (4.10), but about the explanatory significance of the category itself:

> Just because a taxonomy is right to classify a group of organisms as *Canis* and a subgroup as *Canis familiaris* does not show that there is any explanatory significance in treating the former as a genus and the latter as a species. We still need to show that the category itself does explanatory work.
>
> (Devitt 2011c: 167)

Remarks like the earlier, "taxa of higher rank than species do not exist in the same sense as do species" (Eldredge and Cracraft 1980: 327), should, I surmise, be construed charitably as not really being about taxa but rather about the biological significance of the higher ranks into which taxa are placed.

6.7.2 The Explanatory Issue with the Category **Race**

Against this background of realism issues about the Linnaean categories, we can locate our second issue with *Racial Category Realism*: Is the category **Race** really explanatory? A positive answer requires much more than identifying races with subspecies—a problematic identification in any case, as we noted (6.4)—because the category **Subspecies** is in doubt too:

> it is generally accepted that the subspecies category is (at best) a unit of taxonomic convenience. Race theorists conclude that the implications for human race ought to be obvious. If the subspecies concept is inadequate for defining non-human subspecies, we should not use it to divide humans into biological races.
>
> (Andreasen 1998: 205)
>
> Racial naturalism depends not only on the validity of human subspecies classification, but also more generally on the taxonomic validity of the "subspecies" category itself. It is important to note that the subspecies concept has been the focus of considerable debate. (Hochman 2013: 343; see also 2014: 81)[27]

[27] And note this: "in modern systematic practice, the distinction between species and subspecies ranking is almost always based on some subjective measure of the degree of phenotypic (in the broad sense) differentiation" (Cracraft 1983: 99).

Just as the view that the **Species** category is real requires a "concept" that shows how being in that category is explanatory, so too does the view that the **Subspecies** category is real. And so too, the **Ecotype** category. Identifying **Race** with **Subspecies** or **Ecotype** just postpones the problem of producing an explanatory "race concept". *Racial Category Realism* requires a theory that any kind is a race, and so is in the category **Race**, in virtue of having a certain sort of explanatory property. But we should not be too demanding. Thus there is no need to go along with Maglo's demand that race be "a primary or foundational concept in human population genetics" (2010: 362; citing Gannett 2004). **Race** needs to be explanatory not foundational.

There have been several attempts to produce robust race concepts; we have already mentioned some examples, Andreasen (1998), Kitcher (2003), Pigliucci and Kaplan (2003). (And we should see the "definitions" mentioned in section 6.4 as gestures toward such a concept.) Maglo is skeptical of them all. He rejects "various concepts of race, including conceptions of race as a breeding population, continental cluster and ancestral line of descent, or clade" (2011: abs). This leads him to the view that there is *no* way in which **Race** is explanatory.

Let us consider what Maglo describes as "a dominant concept of race in dispute among genomic scientists . . . the continental ancestry identification race concept" (2011: 370). Support for this concept has come particularly from the study by Rosenberg et al. (2002), summarized in subsection 6.5.3. We considered there the bearing of this study on taxon issues but noted that it has primarily been brought to bear on *Racial Category Realism (Humans)*. The realist idea is that for a kind, for example, Eurasians, to be a race is for it be a genetic cluster, particularly at the K = 5 level (Hardimon 2017; Spencer 2019: 99–104). But then the crucial problem is: Why is being such a cluster biologically significant? It is not sufficient to say that it picks out kinds that match up nicely with (at least some of) our pre-theoretical classifications of races. We need to know why being one of those clusters is significant for being a race, rather than arbitrary.

For example, why take K = 5 clusters as races rather than the K = 3 or K = whatever? Pigliucci, responding to Sesardic (2010), states trenchantly that picking K = 5 "is blatant cherry picking of the relevant evidence" (2013: 273). Kitcher points out that "there is a genuine issue about *level* or *fineness of grain*" (2007: 305). Hochman has this to say:

> The appropriate grain of analysis is unclear. Is there one American race, as suggested by a racial reading of the worldwide analysis, or five, the number of clusters identified in the within-continent analysis? . . . The Americas have hundreds of potentially genetically distinguishable populations. (2013: 345)

This could lead to "a *reductio ad absurdum*, as the race naturalist would have to concede that we have hundreds, even thousands of 'races'" (p. 345).

Here's another problem. Hochman claims that "in nonhuman biology single-locus statistics are the standard tools used for subspecies classification" (2013: 342). So why use the method of genetic clustering when it comes to humans? We need "a forceful argument for using genetic clustering, and against using single-locus statistics, for subspecies classification" (pp. 341–2). Genetic clustering and single-locus yield different subspecies concepts. Which is correct? It needs to be argued that it is in virtue of being a kind of the one sort rather than the other that the kind is biologically explanatory. Hochman, in effect, doubts that either of these concepts can be thus supported.

Hochman thinks that "the inference from 'genetic cluster' to 'subspecies,' fundamental to racial naturalism, stands unsupported" (p. 344). The inference may well be unsupported but how "fundamental" is it? Not at all to *Racial Taxon Realism (Humans)*. But the failure of the inference is certainly not helpful to a robust *Racial Category Realism (Humans)*.

I shall return to the continental ancestry concept later, including to its alleged arbitrariness. Meanwhile, let us suppose that this concept cannot sustain realism. And I rather doubt that any of the other proposed race concepts criticized by Maglo can be shown to be robustly explanatory. So, let us suppose that Maglo is right and they cannot. Just how serious would this be?

6.7.3 Minimal Concepts of the Higher Categories

We note first that the lack of a robust concept seems to put **Race** in the same boat as **Genus** and other higher categories; see the widespread doubts about the Linnaean categories aired above (6.7.1).[28] That is to say, we seem to lack robust category concepts for all the higher categories. Indeed, even the category **Species** is a bit problematic because it is blessed with *many competing* robust concepts, leading some to species pluralism. But, despite the widespread doubts about the higher categories, it is very revealing how much biologists keep talking about them.[29]

It is hard to exaggerate just how much biologists do talk about them. Consider, for example, the International Code of Zoological Nomenclature (1999), surely something very important to biological theorizing. It is full of talk of genera, families, etc., starting with this:

[28] Futuyma, in effect notes this: "subspecies are quite arbitrarily defined: and... the same is true of genera, families, and other higher categories of classification" (1986: 111).

[29] My discussion draws on my "The Minimal Role of the Higher Categories in Biology" (2023).

> The Code regulates the names of taxa in the family group, genus group, and species group. Articles 1–4, 7–10, 11.1–11.3, 14, 27, 28 and 32.5.2.5 also regulate names of taxa at ranks above the family group. (1.2.2)

And here is some typical Code talk about the hierarchy:

> The genus group, which is next below the family group and next above the species group in the hierarchy of classification, encompasses all nominal taxa at the ranks of genus and subgenus (see also Articles 10.3 and 10.4). (42.1)

Wikipedia articles about biological taxa are replete with talk of the categories. So too is Sterelny and Griffiths' textbook, including this:

> especially in paleontology, evolutionary patterns are often studied at the level of the genus or family, rather than by identifying individual species.... So when evolutionary theorists writing on the history of life...the information they extract will mostly be patterns of family extinction, survival, or spread. (1999: 194)

Finally, note the talk of a very high category that begins a paper by James Valentine: "It is consistent with fossil evidence that all living metazoan phyla originated during the late Precambrian and Cambrian." Valentine goes on to address the issue of "why there was not a continuing evolutionary innovation at the level of phyla" (1995: 190).

The ubiquitous talk of the higher categories in biology is surely not just idle rhetoric. Ereshefsky thinks that "the widespread acceptance of the Linnaean system" stems from "virtues" that are "just pragmatic and not theoretic", and cites evidence that this is a common view (Ereshefsky 2001: 233). But the view strains credulity. Talk of the categories does, of course, *have* pragmatic benefits, but *so too does the talk of any true theory*: a *test* of truth is success in practice. The apparent commitments of a theory must be taken seriously, as Quine (1960) emphasized; one cannot have one's theoretical cake and eat it too. So, if the virtues of biology's talk of the categories were really "just pragmatic", then we ought to be able to paraphrase the talk away *without theoretical loss*. We have no reason to believe that this is possible; that, for example, Valentine could raise his issue without talking of phyla. Rather, we should accept that this talk of the higher categories is doing some theoretical work and address the question, "What work?" Sterelny and Griffiths mention "the importance of the information expressed" by this talk (1999: 201). We need to consider: What information?

The first step in answering is to emphasize that doubts about the Linnaean hierarchy are not doubts that there is a hierarchy of taxa; "The challenge here is not to the assumption that life is hierarchically arranged" (Ereshefsky 1999: 299).

Canidae (a large group of "dog-like" animals) really does include *Vulpes* ("true foxes"), which includes *Vulpes vulpes* (red foxes). The challenge is to the view that this hierarchy of taxa can be captured by the Linnaean categories.[30] Next, with this very real hierarchy in mind, consider the Code's description of the "genus group" being "below the family group and next above the species group". This typical description points the way to my "minimal" proposal. *If nothing else*, a category is a kind of taxon at a certain rough level in the hierarchy of taxa. As we have already noted in discussing California snakes (6.3), the level of a taxon in the hierarchy has consequences for its explanatory power. So, the categories do have an informational and explanatory role, even if not one as "robust" as biologists who worry about the categories may have hoped for. We *may* have to settle for "minimal" category concepts, according to which Linnaean categories capture, in a somewhat rough and ready way, the explanatorily significant hierarchy of taxa.

Consider a particular red fox, Rufus, a member of the taxon, *Vulpes vulpes*. The fact that Rufus is a member of that taxon explains very many of his phenotypic properties; for example, that he has a bushy tail with a white tip (1.3, 2.3). For, the very nature or "essence" of red foxes causes them to have such tails (in their normal environment); it is *because* of that essence that Rufus has that tail; the essence must meet the Sober demand (2.5). The explanations in question, of course, might be structural or historical.[31]

Moving up a level in the hierarchy, Rufus is also a member of the taxon, *Vulpes*, the taxon of "true foxes". Many of Rufus' phenotypic properties explained by his being a *Vulpes vulpes*, including having that tail, are *not* explained by his being a member of the higher taxon, *Vulpes*; many true foxes, including Arctic foxes (*Vulpes lagopus*), do not have such tails. However, the fact that Rufus is a *Vulpes* does explain *some* of his properties and, importantly, explains those same properties in animals from many other taxa at the lower level. Rufus' property of having partially retractable claws is an example, for Rufus shares this property with Arctic foxes, Cape foxes (*Vulpes chama*), and other true foxes. Their having this property is explained by their being members of the taxon *Vulpes*; the very nature of *Vulpes* causes them to have such claws (in their normal environment); it is *because* of that nature that these various true foxes have those claws.

There is a similar story when we move up another level in the hierarchy of taxa. Rufus is also a member of the taxon, *Canidae*. Many of the properties of true foxes, including their claws, are not explained by their being members of the higher taxon, *Canidae*; dogs and jackals do not have retractable claws. However, the fact that Rufus is a *Canidae* does explain some properties he shares not only with other

[30] Ereshefsky (1999: 299–302) and Mishler (1999: 311–12) discuss other ways to capture the hierarchy.

[31] The species' place in the phylogenetic tree will, of course, be central to the historical explanation.

true foxes but also with members of the taxon *Canis*; for example; the property of walking on toes ("digitigrade"). This property of true foxes, dogs, jackals, and many other taxa is explained by the nature of *Canidae*.

We see here, what we noted before in discussing California snakes (6.3), that the explanatory significance of being a member of a taxon comes in degrees along two dimensions. On one dimension, being a *Vulpes vulpes* explains more of the phenotypic properties of Rufus than does being a *Vulpes*, which in turn explains more of those properties than being a *Canidae*, and so on up the hierarchy. On the other dimension, being a *Canidae* explains *some phenotypic properties of more organisms* than does being a *Vulpes*, which in turn explains more of those properties than being a *Vulpes vulpes*, and so on down the hierarchy. So, a taxon's place in the very real hierarchy of taxa is explanatorily significant.

There has been no talk of categories so far, but now we introduce them: we claim that*Vulpes vulpes* is in the category **Species**, *Vulpes*, in the category **Genus**, *Canidae*, in the category **Family**. This claim conveys, *at least*, information about the relative explanatory power of these taxa. Thus, on the one hand, the claim informs us that, since *Vulpes vulpes* is a species, being in that taxon, explains more of the phenotypic properties of its members, than simply being in their genus, *Vulpes*. And, on the other hand, simply being a *Vulpes* explains some phenotypic properties not only of members of the species *Vulpes vulpes* but also of all the members of other species in the genus. Moving up the hierarchy, we get a similar comparison between the explanatory powers of genus *Vulpes* and family *Canidae*. And so on up.

The level of a taxon in the hierarchy of taxa is an indication of its explanatory power along the two dimensions. According to this minimal concept of a Linnaean category, an ascription of the category to a taxon is an attempt to identify the explanatorily significant level of the taxon, if nothing more. Whereas ascription of a taxon to an organism explains its phenotypic properties, ascription of a category to a taxon explains (at least), the scope of explanations that the taxon can yield. The criticisms of the Linnaean categories show that ascriptions of those categories are *imperfect* guides, providing *imprecise* information about levels. Nonetheless, these ascriptions do provide important information about those levels, as we have just seen.

Other sciences have hierarchies of kinds, of course; for example, chemistry, geology, and astronomy. The same sort of explanatory significance attaches to levels in those hierarchies: a kind *L* that is below kind *H* in a hierarchy explains more of the observable properties of its members than does *H*; but *H* explains not only some observable properties of members of *L* but also those same properties of members of other kinds below *H* in the hierarchy. And with the significance of levels goes an interest in categorizing levels. Consider chemistry, for example. Its situation seems to be interestingly different from that in biology (Djoumbou Feunang et al. 2016). Whereas biology has, sort of, settled on the Linnaean hierarchy of categories, controversial though it is, chemistry has several such hierarchies. But, whereas biology struggles to find definitions for its categories, chemistry seems to have promising definitions available for its; see those accompanying a recently proposed "automated chemical classification" with a hierarchy of 11 levels (Djoumbou Feunang et al. 2016).

Objection. "According to your minimal concepts, the identification of a category is purely relational. The concept for category *K*, identifies *K* as the category of taxa with the appropriate explanatory power relative to taxa in category *J* above and taxa in category *L* below. But how are *J* and *L* identified? Take *J*. It is the category of taxa with the appropriate explanatory power relative to taxa in category *I* above and *K* below. And so on. Categories identified in this purely relational way can provide no information about the explanatory power of any of these taxa." This is a good point. Absent an independent *non*-relational identification of at least one category, this account is fairly empty. We need the explanatory power of taxa in that independently identified category, a category with a robust not minimal concept, to serve as the anchor of this hierarchy of categories; comparison with that category can then yield genuine information about the explanatory power of taxa in other categories. We know immediately where to go for the anchor, the **Species** category; "one taxonomic category, the species, has been thought by many to be real and nonarbitrary" (Futuyma 1986, p. 111). For, there are many robust competing species concepts each of them identifying the **Species** category without adverting to any other category. We already have, of course, a good idea of the explanatory power of the taxa in that category, as those concepts indicate. In any case, the actual explanatory power of species serves as the anchor. Thus, knowing that a taxon is in the category **Genus**, provides information about the taxon's explanatory power given what we know about the explanatory power of taxa in the independently identified category **Species**. And so on up.

So, despite the imperfections of the Linnaean categories, and the lack of robust category concepts above **Species**, the categories are at least explanatory in this minimal way of marking out, *in a somewhat rough and ready way*, a level in the hierarchy of explanatory taxa. The categories are not as explanatorily important as the taxa they categorize but they are indeed informative, as Sterelny and Griffiths noted. The ubiquitous talk of the higher categories in biology is not "just pragmatic"; it serves a theoretical purpose. The boat that the Linnaean categories are in is not *totally* unseaworthy.

If we think of the Linnaean hierarchy as simply an attempt to mark out explanatory levels, its imperfection is not surprising. It can be a tricky matter deciding whether to propose a taxon at a level between those of two taxa, between a taxon that would include the proposed taxon and a taxon that the proposed taxon would include. It may be hard to say whether our explanatory needs justify such a proposal. Indeed, there may be a deal of indeterminacy about the appropriate number of levels. And it is not surprising that decisions about levels in one group of organisms and decisions in another do not match up levels neatly.

Still, the Linnaean categories seem *not* to be simply distinguishing levels but attempting to tell us something about each level. This is brought out vividly by monotypic taxa, taxa that contain only one immediately subordinate taxon. Consider, for example, the beluga (white) whale: one and the same group of animals form the monotypic genus, *Delphinapterus*, and its only species, *Delphinapterus leucas*. We should see our calling that group a genus as an attempt to convey something theoretically important beyond calling it a species. The minimal concept of **Genus** does not remove the need for a robust concept to justify this talk.

6.7.4 A Minimal Concept of the Category **Race**

Return now to the category **Race**. To say that a taxon is a biological race is to say that it is a kind of biological taxon that is at an explanatory level below **Species**. What level? We noted in section 6.4 that 'race', like other names for infraspecific categories, is used "informally and a bit loosely" in biology. As a result, **Race** is identified sometimes with **Subspecies**, which is at the first infraspecific level, but sometimes with **Ecotype or Form**, which are at a lower level. So, the category is undesirably vague and indeterminate: there is no fact of the matter which infraspecific level of taxa is a race.

Still, I emphasize that it is explanatory to know that a taxon is a race, even with the indeterminacy, just as it is to know that it is, say, an ecotype. Thus, consider the Pacific Islanders. The claim that they are a race tells us that being a Pacific Islander explains *more of the phenotypic properties of its members* than does being in the species *Homo sapiens*. But being a member of that species *explains some phenotypic properties of more organisms than does being in that race*, because it explains those properties not only of members of that race but also of all the other races in the species. So, even if Maglo is right and **Race** is not robustly explanatory, it is at least explanatory in this somewhat minimal way of marking out, too vaguely, a level in the infraspecific hierarchy of taxa. **Race** is not as explanatorily important as taxa like Pacific Islanders that it is thought to categorize but it is not unimportant. So, this "minimal" *Racial Category Realism (Humans)* puts **Race** in the same boat as the Linnaean categories and, as noted, that boat is not totally unseaworthy. (We shall return to the indeterminacy issue in section 6.8.)

Dupré's discussion prompts two further comments. First, interestingly, he speaks favorably of the biological significance of "local human 'ecotypes'" which he is at pains to distinguish from races:

> The point is just that while there have been, and continue to be, numerous very local human types adapted to specific local conditions, this is a vastly finer-grained classification than any standard racial category. (2008: 67)

Dupré gives the nice example of Kenyans. They exist in "a culture involving extensive running at high altitude" (pp. 67–8). This may have caused them to evolve in a way that makes them apt to be the good marathon runners that so many of them are. If so, being Kenyan is an explanatory interesting kind. What sort of kind? We have already noted that biologists often posit an infraspecific *hierarchy* along these lines: below **Subspecies** there is **Variety** and **Subvariety**; there is **Form** and **Subform**; and, important for our purposes, there is **Race** and **Strain** (6.4). So, it seems apt to say, following some biologists, that these Kenyans may form a strain of the black African race of the human species.

Second, Dupré continues:

> Broad racial categories, at any rate, comprise large numbers of ecotypes that are likely to differ in most respects of local adaptation. (p. 68)

Right: broader categories "differ in most respects of local adaptation" from finer-grained categories *all the way up the hierarchy*: thus, strains differ from races; following some biologists, races differ from subspecies; subspecies differ from species; species, from genera, and so on. The same concern to explain phenotypic properties that leads us to distinguish species from genera leads us to distinguish races from species and subspecies and strains from races.

So, I am suggesting that racial realists may have to settle for minimal *Racial Category Realism (Humans)*. How disappointing should that be? First, the lack of a robust **Race** concept is clearly disappointing. Still, I suggest, it should be only mildly disappointing. After all, we lack a robust concept for **Genus** and other higher categories but still find it useful to talk about them, as indeed we should, given the minimal concept. The minimal race concept provides a basis for thinking of Amerindians as a race just as the minimal genus concept provides a basis for thinking of *Vulpes* as a genus. Of course, given the noted indeterminacy of **Race**, it marks out an explanatory level in an even more rough and ready way than does **Genus**, but it still is explanatory. Second, even the minimal concept addresses the fineness-of-grain worry. Thus, the above supposition that Kenyans are a strain of the black African race puts Kenyans in a more fine-grained category than Africans, and Africans in a more fine-grained category than their species, humans. And this ranking in the categories is explanatorily significant, as we have demonstrated, even if not as explanatorily significant as one might have hoped.

To allay any disappointment here, the racial realist should keep in mind that these difficulties for *Racial Category Realism* do not reflect on *Racial Taxon Realism*. It remains the case that there are good explanatory grounds in biology for dividing some species, including *Homo sapiens*, into kinds thought to be races. That is surely what should matter most to racial realism. Thus, compare the situation for **Race** with that for **Genus**. It is clearly very important to biologists that they have good grounds for "genus taxon realism": they have good grounds for identifying and naming the likes of *Vulpes*, a taxon thought to be a genus. It is disappointing that biologists lack a robust concept to support "genus category realism", but this lack is apparently not so important, given that biologists keep talking of **Genus** nonetheless. What is good enough for genus realists should be good enough for racial realists.

Nonetheless, as acknowledged, **Race** is undesirably indeterminate. In the face of this, I suggest that the racial realist should push to make **Race** more precise. We have noted that some biologists, at least, put **Race** at a higher level than **Strain** and at a lower level than **Subspecies**. Templeton (2013) has made the identification of **Race** with **Subspecies** seem very unpromising. He makes a persuasive case that there are three subspecies of chimpanzees: "Upper Guinea, Gulf of Guinea, and the combined equatorial African populations" (p. 266). These three taxa have distinct evolutionary lineages and have an F_{st} value for genetic diversity of more than 0.25. The alleged human races do not meet these two criteria for subspecies, as we noted: their F_{st} value is only 0.043 and they form a single evolutionary lineage. In brief, they do not differ nearly as much as the chimpanzees. Indeed:

> Humans show only modest levels of differentiation among populations when compared to other large-bodied mammals, and this level of differentiation is well below the usual threshold used to identify subspecies (races) in nonhuman species. (Templeton 1999: 646)

So the promising path for the racial realist about humans is clearly to remove some of the indeterminacy by ranking **Race** at some level below **Subspecies**. And the realist should go along with biologists who think of races as above strains, a level that might include the likes of Kenyans. Then, races would be at a less fine-grained level than those smaller groups, just as genera are at a less fine-grained level than species. That would be part of what it is to be a race or a genus on the minimal concepts. Then black Africans might count as a race but Kenyans would not.

This move to precisify **Race** raises another question. **Race**, as it stands, is certainly indeterminate but is it also *arbitrary*? We allowed that there may be some arbitrariness about races at the taxon level (6.5) but what about the category level? I think that the category **Race** is not arbitrary. Though there is no fact of the matter about precisely which infraspecific explanatory level is **Race**, there is nothing arbitrary *about those levels*. However, arbitrariness does get into the picture when we move to precisify **Race**. Thus, consider the above recommendation to exclude the identification of **Race** with **Subspecies**. This strengthens the case for treating groups of humans that are commonly thought of as races—for example, Amerindians—as really being races, but that hardly counts as a sound biological reason for the exclusion. The truth must be that it is indeed arbitrary to some extent which level in the hierarchy of infraspecific levels of explanation we decide to label "Race" when we precisify. But this arbitrariness does not strike me as particularly interesting. Classes are above orders in the Linnaean hierarchy. Would it have mattered if classes had been labeled "orders" and orders "classes"?

In light of this, let us revisit the value-of-K problem (6.7.2). The realist idea that came under criticism was that races are genetic clusters picked out by K = 5. One problem was: Why K = 5 not K = 3 or K = whatever? And the answer *may* be: because the kinds picked out by K = 5 are at a sufficiently explanatory infraspecific level. This does not rule out that *other* values of K may yield other kinds at other explanatory levels. As we have been emphasizing, there can be explanatory infraspecific kinds at several levels. If there are with humans, that would raise the question which one, if any, of these values of K picks out *races*. And, given the indeterminacy we have been emphasizing, there is to some extent no fact of the matter about the answer to this question. And stipulating a level to be **Race** is, to some extent, arbitrary. But whether K = n is even a candidate for picking out races depends on the kinds in question being sufficiently explanatory.

This bears on a point made by Kaplan. He notes that "the human species had and has population structure" (2011: 3). Drawing on the work of Novembre et al. (2008), he asks

> Are there many 'races' within Europe? The kinds of 'clustering' that lead some researchers to argue that socially recognised human races have a genetic basis can also be used to sort populations into much smaller groups, groups that no one seriously argues are of any biological significance. (p. 4)

Thus it is possible to distinguish by a genetic basis some European nations from others and even some geographic populations within a nation. As Faucher sums up:

> Genetic clustering can produce many different clusters at many different levels and assign individuals to ethnic groups that do not correspond to races. Why, then, should we privilege continental races more than other clusters at lower or higher levels? (2017: 257)

What matters most in considering any of these genetically based groups is, as Kaplan would say, whether they provide "useful biological information" and so are "biologically significant"; and, as I would say, whether they are sufficiently explanatory in biology. If they are, whether they count as "races" may be indeterminate but is not so important anyway.

Let me sum up my position on the importance of these various realism doctrines. First, the most important doctrine is *Racial Taxon Realism*: the division of organisms into infraspecific kinds thought to be races is biologically important, even if not nearly as important as the division of them into species. Still, second, even minimal *Racial Category Realism* is of some importance. Just as the higher categories are explanatory in marking out, in a rough and ready way, a level in the hierarchy of explanatory taxa, so too is **Race**. But, third, **Race**, as it stands, is undesirably indeterminate. We would get a more important doctrine if we precisifed **Race**, in particular rejecting the identification with **Subspecies**. Fourth, I can see no good reason for denying that **Race**, as it stands or precisified, is properly called 'race'. But whether or not I'm right about that does not strike me as an important issue. We shall return to it in the next section.

I conclude this section by raising the possibility of our *not* having to settle for a minimal race concept. We noted the standard 0.25 criterion of genetic diversity for subspecies. Now the main problem with this, as Templeton notes, is its "arbitrariness" (2013: 263). But suppose that we can show that infraspecific kinds meeting that criterion, *or something like it*, and with different evolutionary histories, exemplify an explanatorily significant level. Then we would have removed the arbitrariness and would have a robust subspecies concept. That could serve as an inspiration for robust lower-level concepts, including one of race. Perhaps we can find a lesser criterion of genetic diversity, and a lesser criterion of evolutionary difference, for explanatory kinds at a lower level that could constitute being in the category at that level.

6.8 Too Weak to be Interesting?

Some, perhaps many, will remain unconvinced by this defense of racial realism, thinking it does not provide enough basis for thinking of, say, Amerindians as a *race*, and that this is fatal to racial realism. Hochman is surely one such. He may respond that my racial realism does not deny anything that anti-realists like him believe:

> It is a misconception that anti-realists about biological race believe that 'race' is totally uncorrelated with any biological difference: we just believe that it does not capture very much biological difference, and that it does not capture that difference very well. (2014: 81)

He may find my realism too weak to be interesting:

> on a weak definition of racial naturalism [my realism] race naturalists wind up saying something profoundly uninteresting—something we knew all along—and hardly anything that anti-realists about biological race would disagree with, apart from the use of the term 'race'. (2014: 81)

Now I agree that there is not "very much biological difference" between human race taxa. Still, there is enough difference to be explanatorily significant; see my discussion of *Racial Taxon Realism (Humans)* in section 6.5.[32] I agree that calling those taxa "races" "does not capture [the level of difference] very well". Still it does capture it a significant amount; see my minimal **Race** concept above (6.7.4). So, where do we disagree?

The most prominent disagreement will be "about the use of the term 'race'". Thus, consider his response to Spencer's (2014) "genetic clusters" view of "racial naturalism":

> The problem is semantic. Race naturalists need to convince the rest of us that their preferred race concept is what we should mean when we talk about race. (Hochman 2014: 84)

He comments similarly on Theodosius Dobzhansky's (1944) view, which he calls "race-as-population", and Pigliucci and Kaplan's (2003) view, which he calls "race-as-ecotype":

> We know that there are human populations and human ecotypes. The controversy is all about the semantics. It is about whether or not 'population' or 'ecotype' is what we ought to mean when we ask, "Are there human races?"... Race-as-population and race-as-ecotype are weak forms of racial naturalism, which need to be defended on semantic grounds. (Hochman 2014: 84)

Hochman thinks that racial naturalism should embrace

> how race was defined by early race theorists, such as Kant.... This definition offers a distinctive and substantive theory about human biological diversity. Other definitions either change the meaning of race so much that they constitute a change in topic or weaken its meaning to an extent that renders racial naturalism trivial and uninteresting. (2019: 1249)

We have another worthy-of-the-name issue. The previous issue was whether our use of 'race' was in order given its ordinary meaning (6.6). The present issue is whether our use of 'race' is in order given a Kant-inspired, explicitly "stipulative", theoretical meaning. I shall not consider this stipulative definition. I think that this venture into normative semantics is a mistake.

First, suppose that we did consider Hochman's stipulative definition and found that my use of 'race' does not comply with it. (It is not obvious to me that we would find this.) Why would that matter? I claim to have shown in sections

[32] Still, there is some difference and that would be too much for at least one antirealist: "Race... captures no biological truth and, at least applied to the human species, does not capture any biological differences" (Sundstrom 2002: 101).

6.4–6.5 that my use of 'race', including its application to humans, is in accord with *actual scientific usage*. There are biologists who call many infraspecific kinds "races" and these are just the sort of kinds that I am being realist about. What could be the basis for a stipulative definition that forbids this usage in science? Still, second, suppose that there was some basis for forbidding the application of 'race' to those infraspecific kinds, how important would that be? Not very. The minimal **Race** concept picks out these kinds, at a roughly determined explanatory level, whether they are called 'race' or whatever. (And it would capture them better if we precisified the concept.) Once again, what's in a name?

One senses a hang-up over a word. Yet, *words really are arbitrary*. If the association of 'race' with racism makes the term too toxic to use in discussing the human species, then perhaps we could contemplate some new term, "X", for these infraspecific kinds. But before long, we can confidently predict, that term's association with X-ism will make it similarly toxic.[33]

In sum, Hochman's "semantic" issue may be socially significant but it is not scientifically so. From a scientific perspective the issue is "merely verbal".

So what *significant scientific* difference do I have with a Hochmanian anti-realist? First, I differ in emphasizing the distinction between the taxon and category doctrine. Second, in claiming that the taxon doctrine is more important and *clearly true*. Third, in arguing for a minimal realism at the category level, a realism that is not so bad, given that it may be all we can have for the higher Linnaean categories. Fourth, presumably, in my stance on what Hochman calls "strong racial naturalism", a doctrine that he does oppose (unlike "weak racial naturalism" which he finds too uninteresting to oppose).

Strong racial naturalism is

> the view that race is a privileged, objective, scientific representation of human biological diversity and population structure, and that there are a relatively small number of human races. (Hochman 2014: 82)

Putative rac*es*—for example, Amerindians—are certainly an "objective, scientific representation of human biological diversity". So too is the category **Race**, even on the minimal concept, albeit a category that is undesirably indeterminate. So far then, my racial realism accords with the strong doctrine that Hochman opposes.

But what about "privileged"? There seem to be several explanatorily significant infraspecific levels of biological diversity. There is no need to claim that any one of them is privileged, any more than there is a need to claim that any one of the higher categories is privileged. There is no need to resist the view that "although

[33] One is reminded of the desperate attempt through history to find euphemisms for a familiar implement in the home: 'water closet'; 'WC'; 'lavatory'; 'latrine'; 'toilet'; 'bathroom'; 'restroom'; 'washroom'; 'ladies' room'; finally, my favorite, 'powder room'.

population substructure occurs at different divisionary levels within our species, there is apparently no objectively natural division that is uniquely useful" (Maglo 2010: 360).

Finally, what about the problem of the number of human races? According to Hochman, a consequence of the "weak" doctrine is that "the numbering of races will be arbitrary" (2014: 82).

> Once we lose the anchor in folk taxonomy, the question of the number of races becomes a problem for biological racial realists, because they must devise a principled way to determine this number.... What justification could there be for calling those [skulls] from racialized groups the "races" and those from much smaller groups (be they nations, tribes, etc.) something else?
>
> (Hochman 2021: 454)

The problem here is not one of arbitrariness but of indeterminacy. Given that the biological level of **Race** is identified rather vaguely, there will be indeterminacy about the number of races: at which explanatory level do we count? Do we, for example, count at a level that has Aboriginal Australians, who seem to have diverged from Southeast Asians about 50,000 years ago (Bergström et al. 2016), as a race? But, as we have insisted (6.5.2), indeterminacy is not arbitrariness: whichever of the few infraspecific levels of "objective...human biological diversity" we choose will yield a number. And the way forward for racial realism, I suggested, is to remove some of the indeterminacy by precisifying the **Race** concept (6.7.4).

Finally, Hochman thinks that it is the task of racial naturalism "to challenge social constructionism" (2013: 350). Well, as I noted at the beginning (6.1). I am not urging a racial realism that opposes social constructivism about some social kinds, like perhaps UK blacks, that may be races in some sense. But clearly my realism does oppose a social constructivism that denies any biological reality in another sense.

6.9 Conclusion

What is "racial realism"? My first step in answering this question is to distinguish, as I so often have in this book, a taxon issue from a category issue. I took both issues to be about whether kinds are explanatory:

> *Racial Taxon Realism*: There are infraspecific kinds, thought to be races, that are biologically explanatory.
>
> *Racial Category Realism*: There are infraspecific kinds that exemplify a biologically explanatory category **Race**. (6.3)

Looking to biology, I argued that *Racial Taxon Realism* is clearly true. There are infraspecific kinds, thought to be races, that are, like the Linnaean taxa, biologically explanatory in virtue of having a part intrinsic and part historical essence (6.4). But what about humans in particular? Are the alleged human races biologically explanatory? Is *Racial Taxon Realism (Humans)* true? I argued that it is (6.5.1). The phenotypic differences between those races seem just the sort of fairly minor differences between members of a species that have prompted biologists to posit the sorts of infraspecific taxa that biologists often call "races".

I argued against four objections (6.5.2): (I) that the differences between alleged races are *not significant enough*, to warrant positing different races; (II) that it is biologically arbitrary to pick out the alleged races from other groups that differ in similarly minor ways; (III) that human variation is clinally distributed; (IV) that the required genetic basis for traditional racial groups is lacking. Two claims were important to my responses. First the claim that biological kinds *will* be a bit indeterminate in that there is no fact of the matter whether or not certain organisms are members of them. Second, the claim that kinds *may* be arbitrary in that there is no persuasive explanatory reason for a theory naming one (indeterminate) kind rather than another that overlaps with it.

Turning to *Racial Category Realism (Humans)*, I first rejected an objection arising from the claim that, in its ordinary meaning, 'race' refers to social kinds (6.6.). The objection is then that the alleged biological races are not "worthy of the name" 'race'. I argued that this claim rests on questionable semantic assumptions and is not obviously true. Even if it is true, the use of 'race' in biology is, at least, related to an ordinary use. Finally, I wondered what's in a name anyway.

I then considered the more serious objection that the category **Race** is *not* explanatory (6.7). I emphasized that if this is true it seems to put **Race** in the same boat as the higher Linnaean categories. I argue that though all of these categories may indeed not be "robustly" explanatory, they are "minimally" so. They are explanatory in marking out, *in a rough and ready way*, levels in the hierarchy of explanatory taxa.

Any disappointment about this should be allayed by the truth of *Racial Taxon Realism*. There remain good explanatory grounds in biology for dividing some species, including *Homo sapiens*, into kinds thought to be races. That is surely what should matter most to racial realism.

Finally, I rejected the criticism, likely to come from Hochman in particular, that the racial realism I have defended is too weak to be interesting (6.8).[34]

[34] My thanks to Adam Hochman, Antonella Mallozzi, and Raj Nanavati for helpful comments on a draft of this chapter.

Glossary of Named Doctrines

The biological taxa referred to in these doctrines are the ones thought to fall under the categories in the Linnaean hierarchy (1.1, 2.2).

Taxon Essentialism

Partly Intrinsic Taxon Essentialism (*Intrinsic Biological Essentialism* in ch. 1): Biological taxa have essences that are partly intrinsic underlying, probably largely genetic, properties. (1.1, 2.1)

Partly Historical Taxon Essentialism: Biological taxa have essences that are partly historical properties. (3.1)

Fully Historical Taxon Essentialism: Biological taxa have essences that are not intrinsic but wholly relational, particularly, historical properties. (3.1)

Taxon Essentialism: The essence of a biological taxon is its members' property of having a certain intrinsic underlying property as a result of descent from members of a certain different taxon, where that taxon has a certain different intrinsic underlying component to its essence. (3.7)

Individual Essentialism

Partly Intrinsic Individual Essentialism: Biological individuals have essences that are partly intrinsic underlying, probably largely genetic, properties. (4.1)

Partly Historical Individual Essentialism: Biological individuals have essences that are partly historical properties. (4.1)

Essential Membership: If an individual organism belongs to a taxon it does so essentially. (4.1)

Type Specimens

Levine's Thesis (*de dicto* reading): Necessarily, any species with a type specimen contains its type specimen. (5.1)

De re reading: Any species with a type specimen necessarily contains its type specimen. (5.1)

Race

Racial Taxon Realism: There are infraspecific kinds, thought to be races, that are biologically explanatory. (6.3)

Racial Taxon Realism (Humans): There are infraspecific kinds of humans, thought to be races, that are biologically explanatory. (6.3)

Racial Category Realism: There are infraspecific kinds that exemplify a biologically explanatory category **Race**. (6.3)

Racial Category Realism (Humans): There are infraspecific kinds of humans that exemplify a biologically explanatory category **Race**. (6.3)

References

Andreasen, R. O. 1998. "A New Perspective on the Race Debate". *The British Journal for the Philosophy of Science*, 49: 199–225.

Andreasen, R. O. 2005. "The Meaning of 'Race': Folk Conceptions and the New Biology of Race". *The Journal of Philosophy*, 102(2): 94–106.

Andrews, Kimberly R., Ashley J. Williams, Iria Fernandez-Silva, Stephen J. Newmang, Joshua M. Copus, Corey B. Wakefield, John E. Randall, and Brian W. Bowen. 2016. "Phylogeny of Deepwater Snappers (Genus *Etelis*) Reveals a Cryptic Species Pair in the Indo-Pacific and Pleistocene Invasion of the Atlantic". *Molecular Phylogenetics and Evolution*, 100: 361–71.

Appiah, Kwame Anthony. 1996. *Color Conscious: The Political Morality of Race*. Princeton: Princeton University Press.

Ariew, André. 2003. "Ernst Mayr's 'Ultimate/Proximate' Distinction Reconsidered and Reconstructed". *Biology and Philosophy*, 18: 553–65.

Armstrong, D. M. 1978. *Nominalism and Realism: Universals and Scientific Realism Volume I*. New York: Cambridge University Press.

Austin, C. J. 2017. "Aristotelian Essentialism: Essence in the Age of Evolution". *Synthese*, 194: 2539–56.

Austin, C. J. 2019. *Essence in the Age of Evolution: A New Theory of Natural Kinds*. New York: Routledge.

Bamshad, Mike. 2005. "Genetic Influences on Health: Does Race Matter?" *Journal of the American Medical Association*, 294: 937–46.

Barker, Matthew J. 2010. "Species Intrinsicalism". *Philosophy of Science*, 77: 73–91.

Barry, S. J., and M. R. Jennings. 1998. "*Coluber infernalis* Blainville, 1835 and *Eutaenia sirtalis tetrataenia* Cope in Yarrow, 1875 (currently *Thamnophis sirtalis infernalis* and *T. s. tetrataenia*; Reptilia, Squamata): Proposed Conservation of the Subspecific Names by the Designation of a Neotype for *T. s. infernalis*". *Bulletin of Zoological Nomenclature*, 55(4): 224–8.

Barry, S. J., M. R. Jennings, and H. M. Smith. 1996. "Current Subspecific Names for Western *Thamnophis sirtalis*". *Herpetological Review*, 27(4): 172–3.

Bateson, P. P. G. 1982. "Preferences for Cousins in Japanese Quail". *Nature*, 295: 236–7.

Baum, David A. 2009. "Species as Ranked Taxa". *Systematic Biology*, 58(1): 74–86.

Bergström, Anders, Nano Nagle, Yuan Chen, Shane McCarthy, Martin O. Pollard, Qasim Ayub, Stephen Wilcox, Leah Wilcox, Roland A. H. van Oorschot, Peter McAllister, Lesley Williams, Yali Xue, R. John Mitchell, and Chris Tyler-Smith. 2016. "Deep Roots for Aboriginal Australian Y Chromosomes". *Current Biology*, 26: 809–13.

Biello, David. 2007. "Life's Bar Code: Genetic Tests Unveil 15 New Species of North American Birds". *Science and Technology at Scientific American.com*, February 22. https://www.scientificamerican.com/article/barcode-of-life-genetic-tests-15-new-bird-species/.

Bolnik, Deborah A. 2008. "Individual Ancestry Inference and the Reification of Race as a Biological Phenomenon". In Barbara A. Koenig, Sandra Soo-Jin Lee, and Sarah

S. Richardson (eds.), *Revisiting Race in a Genomic Age*. New Brunswick: Rutgers University Press: 70–85.

Boulter, S. J. 2012. "Can Evolutionary Biology do without Aristotelian Essentialism?" *Royal Institute of Philosophy Supplement*, 70: 83–103.

Boundy, J., and D. A. Rossman. 1995. "Allocation and Status of the Garter Snake Names *Coluber infernalis* Blainville, *Eutaenia sirtalis tetrataenia* Cope and *Eutaenia imperialis* Coues and Yarrow". *Copeia*, 1: 236–40.

Boyd, Richard. 1999. "Homeostasis, Species, and Higher Taxa". In Robert A. Wilson (ed.), *Species: New Interdisciplinary Essays*. Cambridge, MA: MIT Press: 141–85.

Brigandt, Ingo. 2009. "Natural Kinds in Evolution and Systematics: Metaphysical and Epistemological Considerations". *Acta Biotheoretica*, 57: 77–97. https://doi.org/10.1007/s10441-008-9056-7.

Brzozowski, Jerzy A. 2020. "Biological Taxon Names are Descriptive Names". *History and Philosophy of the Life Sciences*, 42: 29. https://doi.org/10.1007/s40656-020-00322-1.

Calcott, Brett. 2009. "Lineage Explanations: Explaining How Biological Mechanisms Change". *British Journal for the Philosophy of Science*, 60: 51–78.

Camargo, Arley, and Jack Jr. Sites. 2013. "Species Delimitation: A Decade After the Renaissance". In Igor Pavlinov (ed.), *The Species Problem: Ongoing Issues*. Rijeka: InTech: 225–47. http://dx.doi.org/10.5772/52664.

Caplan, Arthur L. 1980. "Have Species Become Declasse?" *PSA*, 1: 71–82.

Carroll, Sean B. 2005. *Endless Forms Most Beautiful: The New Science of Evo Devo and the Making of the Animal Kingdom*. New York: W. W. Norton.

Cartright, Nancy. 1983. *How the Laws of Physics Lie*. Oxford: Clarendon Press.

Cavalli-Sforza, L. 2005. "The Human Genome Diversity Project: Past, Present, and Future". *Nature Genetics*, 6: 333–40.

Cavalli-Sforza, L., P. Menozzi, and A. Piazza. 1994. *The History and Geography of Human Genes*. Princeton: Princeton University Press.

Cooper, Richard S. 2003. "Race, Genes, and Health—New Wine in Old Bottles?" *International Journal of Epidemiology*, 32: 23–5. https://doi.org/10.1093/ije/dyg036.

Cope, E. D. in Yarrow, H. C. 1875. "Report upon the Collections of Batrachians and Reptiles made in Portions of Nevada, Utah, California, Colorado, New Mexico, and Arizona, during the Years 1871, 1872, 1873 and 1874": 509–84. In Engineer Dept, USA (ed.), "Report upon Geographical and Geological Explorations and Surveys West of the One Hundredth Meridian", vol 5 (*Zoology*), part 4: 546.

Cracraft, Joel. 1983. "Species Concepts and Speciation Analysis". In R. Johnston (ed.), *Current Ornithology*. New York: Plenum Press: 159–87. Reprinted in Marc Ereshefsky (ed.), *The Units of Evolution: Essays on the Nature of Species*. Cambridge, MA: MIT Press: 93–120. [Citations are to Ereshefsky.]

Cracraft, Joel, Julie Feinstein, Jeffrey Vaughn, and Kathleen Helm-Bychowski. 1998. "Sorting out Tigers (*Panthera tigris*): Mitochondrial Sequences, Nuclear Inserts, Systematics, and Conservation Genetics". *Animal Conservation*, 1(2): 139–50. https://doi.org/10.1111/j.1469-1795.1998.tb00021.x.

Cruse, D. A. 1992. "Review Article: Monosemy vs Polysemy". *Linguistics*, 30: 577–99.

Darwin, Charles. 1859. *The Origin of Species by Means of Natural Selection*. 1st edn 1859. New York: Barnes & Noble Classics (2004).

Davenport, C. B. 1930. "The Mingling of Races". In E. V. Cowdry (ed.), *Human Biology and Racial Welfare*. New York: Paul B. Hoeber: 553–65.

de Blainville, H. 1835. "Description de quelques espèces de reptiles de la Californie, précédée de l'analyse d'un système général d'erpétologie et d'amphibiologie". *Nouvelles Annales du Muséum d'Histoire Naturelle*, 3(4): 291.

De Queiroz, K. 1992. "Phylogenetic Definitions and Taxonomic Philosophy". *Biology and Philosophy*, 7: 295–313.

De Queiroz, K. 1995. "The Definitions of Species and Clade Names: A Reply to Ghiselin". *Biology and Philosophy*, 10: 223–8.

Dennett, Daniel C. 2017. "Darwin and the Overdue Demise of Essentialism". In David Livingstone Smith (ed.), *How Biology Shapes Philosophy: New Foundations for Naturalism*. Cambridge: Cambridge University Press: 9–22.

Devitt, Michael. 1974. "Singular Terms". *Journal of Philosophy*, 71(7): 183–205.

Devitt, Michael. 1981a. *Designation*. New York: Columbia University Press.

Devitt, Michael. 1981b. "Donnellan's Distinction". In Peter A. French, Theodore E. Uehling Jr., and Howard K. Wettstein (eds.), *Midwest Studies in Philosophy, Volume VI: The Foundations of Analytic Philosophy*. Minneapolis: University of Minnesota Press: 511–24.

Devitt, Michael. 1997. *Realism and Truth*. 2nd edn revised "with a new afterword" (1st edn 1984; 2nd edn 1991). Princeton: Princeton University Press.

Devitt, Michael. 2001. "Incommensurability and the Priority of Metaphysics". In P. Hoyningen-Huene and H. Sankey (eds.), *Incommensurability and Related Matters*. Dordrecht: Kluwer Academic Publishers: 143–57. Reprinted in Devitt (2010), *Putting Metaphysics First: Essays on Metaphysics and Epistemology*. Oxford: Oxford University Press: 99–120.

Devitt, Michael. 2005. "Rigid Application". *Philosophical Studies*, 125: 139–65.

Devitt, Michael. 2006. "Intuitions". In Victor Gomez Pin, Jose Ignacio Galparaso, and Gotzon Arrizabalaga (eds.), *Ontology Studies Cuadernos de Ontologia: Proceedings of VI International Ontology Congress (San Sebastian 2004)*. San Sebastian: Universidad del Pais Vasco: 169–76. Reprinted in Devitt (2010), *Putting Metaphysics First: Essays on Metaphysics and Epistemology*. Oxford: Oxford University Press: 292–302.

Devitt, Michael. 2008. "Resurrecting Biological Essentialism". *Philosophy of Science*, 75: 344–82. Reprinted with some substantive additional footnotes in Devitt (2010), *Putting Metaphysics First: Essays on Metaphysics and Epistemology*. Oxford: Oxford University Press: 213–49.

Devitt, Michael. 2009a. "Biological Realisms". In Heather Dyke (ed.), *From Truth to Reality: New Essays in Logic and Metaphysics*. New York: Routledge: 43–65.

Devitt, Michael. 2009b. "On Determining What There Isn't". In Dominic Murphy and Michael A. Bishop (eds.), *Stich and His Critics*. Oxford: Wiley-Blackwell: 46–61. Reprinted in Devitt (2010), *Putting Metaphysics First: Essays on Metaphysics and Epistemology*. Oxford: Oxford University Press: 301–20.

Devitt, Michael. 2011a. "No Place for the A Priori". In Michael J. Shaffer and Michael L. Veber (eds.), *What Place for the A Priori?* Chicago and La Salle: Open Court Publishing Company: 9–32. Also in Devitt (2010), *Putting Metaphysics First: Essays on Metaphysics and Epistemology*. Oxford: Oxford University Press: 271–91.

Devitt, Michael. 2011b. "Experimental Semantics". *Philosophy and Phenomenological Research*, 82: 418–35. https://doi.org/ppr201182222.

Devitt, Michael. 2011c. "Natural Kinds and Biological Realisms". In Joseph Keim Campbell, Michael O'Rourke, and Matthew Slater (eds.), *Carving Nature at its Joints: Natural Kinds in Metaphysics and Science*. Cambridge, MA: MIT Press: 155–73. Also in Devitt (2010), *Putting Metaphysics First: Essays on Metaphysics and Epistemology*. Oxford: Oxford University Press: 197–212.

Devitt, Michael. 2012a. "Whither Experimental Semantics?" *Theoria*, 27: 5–36.

Devitt, Michael. 2012b. "Semantic Epistemology: Response to Machery". *Theoria*, 27: 229–33. https://doi.org/theoria20122711101387/theoria6225.

Devitt, Michael. 2015a. "Testing Theories of Reference". In J. Haukioja (ed.), *Advances in Experimental Philosophy of Language*. London: Bloomsbury Academic: 31–63.
Devitt, Michael. 2015b. "Should Proper Names Still Seem so Problematic?" In Andrea Bianchi (ed.), *On Reference*. Oxford: Oxford University Press: 108–43.
Devitt, Michael. 2018a. "Historical Biological Essentialism". *Studies in History and Philosophy of Biological and Biomedical Sciences*, 71: 1–7. https://doi.org/10.1016/j.shpsc.2018.05.004.s.
Devitt, Michael. 2018b. "Individual Essentialism in Biology". *Biology and Philosophy*, 33: 1–22. https://doi.org/10.1007/s10539-018-9651-1.
Devitt, Michael. 2020. "Stirring the Possum: Responses to the Bianchi Papers". In Andrea Bianchi (ed.), *Language and Reality from a Naturalistic Perspective: Themes from Michael Devitt*. Cham: Springer: 371–455.
Devitt, Michael. 2021a. "Defending Intrinsic Biological Essentialism". *Philosophy of Science*, 88: 1–16. https://doi.org/10.1086/710029.
Devitt, Michael. 2021b. *Overlooking Conventions: The Trouble with Linguistic Pragmatism*. Cham: Springer.
Devitt, Michael. 2021c. "Gogoeta batzuk *Naming and Necessity* buruz". Translation of "Reflections on *Naming and Necessity*" into Basque by Kepa Korta. *Gogoa*, 22: 63–92. https://doi.org/10.1387/gogoa.22599. English version, *Theoria*, special issue on *Naming and Necessity*, 88 (2022): 406–33. https://doi.org/10.1111/theo.12372. [Citation is to the English version.]
Devitt, Michael. 2023. "The Minimal Role of the Higher Categories in Biology". *Biology and Philosophy*, 38(2). https://doi.org/10.1007/s10539-022-09888-511
Devitt, Michael, and Brian Cross Porter. 2021. "Testing the Reference of Biological Kind Terms". *Cognitive Science*, 45. https://doi.org/10.1111/cogs.12979.
Devitt, Michael, and Kim Sterelny. 1999. *Language and Reality: An Introduction to the Philosophy of Language*. 2nd edn (1st edn 1987). Cambridge, MA: MIT Press.
Dickie, Imogen. 2011. "How Proper Names Refer". *Proceedings of the Aristotelian Society*, 101: 43–78.
Djoumbou Feunang, Yannick, Roman Eisner, Craig Knox, Leonid Chepelev, Janna Hastings, Gareth Owen, Eoin Fahy, Christoph Steinbeck, Shankar Subramanian, Evan Bolton, Russell Greiner, and David S. Wishart. 2016. "ClassyFire: Automated Chemical Classification with a Comprehensive, Computable Taxonomy". *Journal of Cheminformatics*, 8: 61. https://doi.org/10.1186/s13321-016-0174-y.
Dobzhansky, T. 1944. "On Species and Races of Living and Fossil Man". *American Journal of Physical Anthropology*, 2: 251–65.
Dumsday, Travis. 2012. "A New Argument for Intrinsic Biological Essentialism". *The Philosophical Quarterly*, 62(248): 486–504. https://doi.org/10.1111/j.1467–9213.2012.00048.x.
Dupré, John. 1981. "Natural Kinds and Biological Taxa". *Philosophical Review*, 90: 66–90.
Dupré, John. 1993. *The Disorder of Things: Metaphysical Foundations of the Disunity of Science*. Cambridge, MA: Harvard University Press.
Dupré, John. 1999. "On the Impossibility of a Monistic Account of Species". In Robert A. Wilson (ed.), *Species: New Interdisciplinary Essays*. Cambridge, MA: MIT Press: 3–22.
Dupré, John. 2002. *Humans and Other Animals*. Oxford: Clarendon Press.
Dupré, John. 2008. "What Genes Are, and Why There Are No 'Genes for Race'". In B. A. Koenig, S. Lee, and S. Richardson (eds.), *Revisiting Race in a Genomic Age*. New Brunswick: Rutgers University Press: 50–74.
Dupré, John. 2017. "A Postgenomic Perspective on Sex and Gender". In David Livingstone Smith (ed.), *How Biology Shapes Philosophy: New Foundations for Naturalism*. Cambridge: Cambridge University Press: 227–46.

Elder, C. L. 2008. "Biological Species *Are* Natural Kinds". *The Southern Journal of Philosophy*, 16: 339–62.
Eldredge, Niles, and Joel Cracraft. 1980. *Phylogenetic Patterns and the Evolutionary Process*. New York: Columbia University Press.
Eldredge, Niles, and Stephen J. Gould. 1972. "Punctuated Equilibria: An Alternative to Phyletic Gradualism". In T. J. M Schopf (ed.), *Models in Paleobiology*. San Francisco: Freeman, Cooper, & Co: 82–115.
Ereshefsky, Marc. 1991. "Species, Higher Taxa, and the Units of Evolution". *Philosophy of Science*, 58: 84–101. Reprinted in Ereshefsky (1992a), *The Units of Evolution: Essays on the Nature of Species*. Cambridge, MA: MIT Press: 381–98. [Citations are to Ereshefsky 1992a.]
Ereshefsky, Marc, ed. 1992a. *The Units of Evolution: Essays on the Nature of Species*. Cambridge, MA: MIT Press.
Ereshefsky, Marc. 1992b. "Introduction". In Ereshefsky (1992a), *The Units of Evolution: Essays on the Nature of Species*. Cambridge, MA: MIT Press: xiii–xvii.
Ereshefsky, Marc. 1992c. "Introduction to Part II: Philosophical Issues". In Ereshefsky (1992a), *The Units of Evolution: Essays on the Nature of Species*. Cambridge, MA: MIT Press: 187–98.
Ereshefsky, Marc. 1998. "Species Pluralism and Anti-Realism". *Philosophy of Science*, 65: 103–20.
Ereshefsky, Marc. 1999. "Species and the Linnaean Hierarchy". In Robert A. Wilson (ed.), *Species: New Interdisciplinary Essays*. Cambridge, MA: MIT Press: 285–305.
Ereshefsky, Marc. 2001. *The Poverty of the Linnaean Hierarchy: A Philosophical Study of Biological Taxonomy*. Cambridge: Cambridge University Press.
Ereshefsky, Marc. 2010. "What's Wrong with the New Biological Essentialism". *Philosophy of Science*, 77: 674–85.
Ereshefsky, Marc. 2014. "Species, Historicity, and Path Dependency". *Philosophy of Science*, 81: 1–13.
Ereshefsky, Marc, and Mohan Matthen. 2005. "Taxonomy, Polymorphism, and History: An Introduction to Population Structure Theory". *Philosophy of Science*, 72: 1–21. https://doi.org/10.1086/426848.
Ereshefsky, Marc, and Thomas A. C. Reydon. 2015. "Scientific Kinds". *Philosophical Studies*, 172: 969–86.
Evans, Gareth. 1973. "The Causal Theory of Names". *Proceedings of the Aristotelian Society*, Supplementary Volume, 47: 187–208.
Faucher, Luc. 2017. "Biophilosophy of Race". In David Livingstone Smith (ed.), *How Biology Shapes Philosophy: New Foundations for Naturalism*. Cambridge: Cambridge University Press: 247–75.
Field, Hartry. 1973. "Theory Change and the Indeterminacy of Reference". *Journal of Philosophy*, 70: 462–81.
Fine, Arthur. 1975. "How to Compare Theories: Reference and Change". *Noûs*, 9: 17–32.
Fine, Kit. 1994. "Essence and Modality". In J. E. Tomberlin (ed.), *Philosophical Perspectives Volume 8: Logic and Language*. Atascadero, CA: Ridgeview Publishing Company: 1–16.
Forbes, G. 1986. "In Defense of Absolute Essentialism". In J. French, T. Uehling Jr., and H. Wettstein (eds.), *Midwest Studies in Philosophy Volume XI: Studies in Essentialism*. Minneapolis: University of Minnesota Press: 3–31.
Futuyma, D. J. 1986. *Evolutionary Biology*. 2nd edn. Sunderland, MA: Sinauer Associates, Inc.
Gannett, L. 2004. "The Biological Reification of Race". *British Journal for the Philosophy of Science*, 55: 323–45.

Gelman, Susan A. 2003. *The Essential Child: Origins of Essentialism in Everyday Thought*. Oxford: Oxford University Press.
Ghiselin, Michael T. 1966. "On Psychologism in the Logic of Taxonomic Controversies". *Systematic Zoology* 26: 207–15.
Ghiselin, Michael T. 1974. "A Radical Solution to the Species Problem". *Systematic Zoology*, 47: 350–83. Reprinted in Ereshefsky (1992a), *The Units of Evolution: Essays on the Nature of Species*. Cambridge, MA: MIT Press: 279–291. [Citations are to Ereshefsky.]
Ghiselin, Michael T. 1987. "Species Concepts, Individuality, and Objectivity". *Biology and Philosophy*, 2: 127–43. Reprinted in Ereshefsky (1992a), *The Units of Evolution: Essays on the Nature of Species*. Cambridge, MA: MIT Press: 363–380. [Citations are to Ereshefsky.]
Ghiselin, Michael T. 1997. *Metaphysics and the Origin of Species*. Albany: SUNY Press.
Glasgow, Joshua M. 2003. "On the New Biology of Race". *The Journal of Philosophy*, 100: 456–74.
Glasgow, Joshua M. 2009. *A Theory of Race*. New York: Routledge.
Godfrey-Smith, Peter. 2014. *Philosophy of Biology*. Princeton: Princeton University Press.
Godman, Marion, Antonella Mallozzi, and David Papineau. 2020. "Essential Properties Are Super-Explanatory: Taming Metaphysical Modality". *Journal of the American Philosophical Association*, 6: 316–34.
Godman, Marion, and David Papineau. 2020. "Species have Historical not Intrinsic Essences". In Andrea Bianchi (ed.), *Language and Reality from a Naturalistic Perspective: Themes from Michael Devitt*. Cham: Springer: 355–65.
Gould, Stephen Jay. 1981. *The Panda's Thumb: More Reflections in Natural History*. Cambridge, MA: Harvard University Press.
Greenberg, D. Robert. 2004. *Language and Identity in the Balkans: Serbo-Croatian and its Disintegration*. Oxford: Oxford University Press.
Greenspan, R. J. 2001. "The Flexible Genome". *Nature Reviews Genetics*, 2: 383–7.
Griffiths, Graham C. D. 1974. "On the Foundations of Biological Systematics". *Acta Biotheoretica*, 23: 85–131.
Griffiths, Paul. 1999. "Squaring the Circle: Natural Kinds with Historical Essences". In Robert A. Wilson (ed.), *Species: New Interdisciplinary Essays*. Cambridge, MA: MIT Press: 209–28.
Griffiths, Paul. 2002. "What is Innateness?" *Monist*, 85: 70–85.
Haber, M. H. 2012. "How to Misidentify a Type Specimen". *Biology and Philosophy*, 27: 767–84.
Hacking, Ian. 2005. "Why Race Still Matters". *Daedalus*, 134(1): 102–16.
Hardimon, M. 2017. *Rethinking Race: The Case for Deflationary Realism*. Cambridge, MA: Harvard University Press.
Haslanger, Sally. 2012. *Resisting Reality: Social Construction and Social Critique*. Oxford: Oxford University Press.
Heaton, T. H. 2016. "The Oligocene Rodent *Ischyromys* of the Great Plains: Replacement Mistaken for Anagenesis". *Journal of Paleontology*, 67(2): 297–308.
Hendry, Robin. Forthcoming. "How (not) to Argue for Microstructural Essentialism".
Hennig, Willi. 1966. *Phylogenetic Systematics*. Urbana, IL: University of Illinois Press.
Hey, Jody. 2001. *Genes, Categories, and Species: The Evolutionary and Cognitive Causes of the Species Problem*. New York: Oxford University Press.
Hochman, Adam. 2013. "Against the New Racial Naturalism". *Journal of Philosophy*, 110(6): 331–51.
Hochman, Adam. 2014. "Unnaturalised Racial Naturalism". *Studies in History and Philosophy of Biological and Biomedical Sciences*, 46: 79–87.

Hochman, Adam. 2019. "Racialization: A Defense of the Concept". *Ethnic and Racial Studies*, 42(8): 1245–62. https://doi.org/10.1080/01419870.2018.1527937.

Hochman, Adam. 2021. "Janus-Faced Race: Is Race Biological, Social, or Mythical?" *American Journal of Physical Anthropology*, 175(2): 453–64. https://doi.org/10.1002/ajpa.24169.

Hull, David L. 1965. "The Effects of Essentialism on Taxonomy: Two Thousand Years of Stasis". *British Journal for the Philosophy of Science*, 15: 314–26; 16: 1–18. Reprinted in Ereshefsky (1992a), *The Units of Evolution: Essays on the Nature of Species*. Cambridge, MA: MIT Press: 199–225. [Citations are to Ereshefsky.]

Hull, David L. 1978. "A Matter of Individuality". *Philosophy of Science*, 45: 335–60. Reprinted in Ereshefsky (1992a), *The Units of Evolution: Essays on the Nature of Species*. Cambridge, MA: MIT Press: 293–316.

Hull, David L. 1982. "Exemplars and Scientific Change". In P. D. Asquith and T. Nickles (eds.), *PSA 1982, Vol. II*. East Lansing: Philosophy of Science Association: 479–503.

ICZN (International Commission on Zoological Nomenclature). 1999. International Code of Zoological Nomenclature. 4th edn. The International Trust for Zoological Nomenclature. http://www.nhm.ac.uk/hosted-sites/iczn/code/.

ICZN (International Commission on Zoological Nomenclature). 2000. "Opinion 1961: *Coluber infernalis* Blainville, 1835 and *Eutaenia sirtalis tetrataenia* Cope in Yarrow, 1875 (currently *Thamnophis sirtalis infernalis* and *T. s. tetrataenia*; reptilia, serpentes): Subspecific Names Conserved by the Designation of a Neotype for *T. s. infernalis*". *Bulletin of Zoological Nomenclature*, 57(3): 191–2.

Jutronić, Dunja. 2010. *Spliski Govor. Od Vapora do Trajekta* [Split Dialect: From Steamboat to Ferryboat]. Split: Naklada Bošković.

Jutronić, Dunja and Filip Galović. Forthcoming. "Govor mjesta Sutivana na Braču" ["Dialect of Sutivan on the Island of Brač"]. Zagreb: Catholic University in Zagreb.

Juvshik, Tim. 2021. "Function Essentialism about Artifacts". *Philosophical Studies*, 178: 2943–2964. https://doi.org/10.1007/s11098-020-01594-w.

Kaplan, J. M. 2011. "Race: What Biology Can Tell Us About a Social Construct". *Encyclopedia of Life Sciences*. Hoboken, NJ: Wiley. https://doi.org/10.1002/9780470015902.a0005857.

Karter, Andrew J. 2003. "Commentary: Race, Genetics, and Disease—in Search of a Middle Ground". *International Journal of Epidemiology*, 32: 26–8. https://doi.org/10.1093/ije/dyg033.

Keil, Frank C. 1989. *Concepts, Kinds, and Cognitive Development*. Cambridge, MA: MIT Press.

Kimbel, W. H., C. A Lockwood, C. V. Ward, M. G. Leakey, Y. Rak, and D. C. Johanson. 2006. "Was *Australopithecus Anamensis* Ancestral to *A. Afarensis*? A Case of Anagenesis in the Hominin Fossil Record". *Journal of Human Evolution*, 51(2): 134–52.

King, Robert C., and William D. Stansfield. 1990. *A Dictionary of Genetics*. New York: Oxford University Press.

Kirk, P. M., P. F. Cannon, D. W. Minter, and J. A. Stalpers. 2008. "Taxonomy". In Kirk et al. (eds.), *Dictionary of the Fungi*. 10th edn. Wallingford: CABI Publishing.

Kitcher, Philip. 1984. "Species". *Philosophy of Science*, 51: 308–33. Reprinted in Kitcher (2003), *In Mendel's Mirror: Philosophical Reflections on Biology*. New York: Oxford University Press: 113–34. Reprinted in Ereshefsky (1992a), *The Units of Evolution: Essays on the Nature of Species*. Cambridge, MA: MIT Press: 317–41. [Citations are to Kitcher 2003.]

Kitcher, Philip. 2003. *In Mendel's Mirror: Philosophical Reflections on Biology*. New York: Oxford University Press.

Kitcher, Philip. 2007. "Does 'Race' Have a Future?" *Philosophy & Public Affairs*, 35(4): 293–317.

Kitts, David B., and David J. Kitts. 1979. "Biological Species as Natural Kinds". *Philosophy of Science*, 46: 613–22.

Kornblith, Hilary. 1993. *Inductive Inference and Its Inductive Ground: An Essay in Naturalistic Epistemology*. Cambridge, MA: MIT Press.

Kripke, Saul A. 1979. "Speaker's Reference and Semantic Reference". In Peter A. French, Theodore E. Uehling Jr., and Howard K. Wettstein (eds.), *Contemporary Perspectives in the Philosophy of Language*. Minneapolis: University of Minnesota Press: 6–27.

Kripke, Saul A. 1980. *Naming and Necessity*. Cambridge, MA: Harvard University Press.

Lackie, John, ed. 2007. *Chambers Dictionary of Science and Technology*. Edinburgh: Chambers Harrap.

Lange, Marc. 1995. "Are There Natural Laws Concerning Particular Biological Species?" *Journal of Philosophy*, 92: 430–51.

LaPorte, Joseph. 1997. "Essential Membership". *Philosophy of Science*, 64: 96–112.

LaPorte, Joseph. 2000. "Rigidity and Kind". *Philosophical Studies*, 97: 293–316.

LaPorte, Joseph. 2003. "Does a Type Specimen Necessarily or Contingently Belong to Its Species?" *Biology and Philosophy*, 18: 583–8.

LaPorte, Joseph. 2004. *Natural Kinds and Conceptual Change*. Cambridge: Cambridge University Press.

Leslie, Sarah-Jane. 2013. "Essence and Natural Kinds: When Science Meets Preschooler Intuition". In Tamar Szabó Gendler and John Hawthorne (eds.), *Oxford Studies in Epistemology Volume 4*. Oxford: Oxford University Press: 108–65.

Levine, Alex. 2001. "Individualism, Type Specimens, and the Scrutability of Species Membership". *Biology and Philosophy*, 16: 325–38.

Lewens, Tim. 2012. "Species, Essence and Explanation". *Studies in History and Philosophy of Biological and Biomedical Sciences*, 43: 751–7.

Lewontin, R. C. 1972. "The Apportionment of Human Diversity". *Evolutionary Biology*, 6: 381–98.

Lisac, Josip. 2009. *Hrvatska dijalektologija 2. Čakavsko narječje* [Croatian Dialectology 2. Chakavian Dialect]. Zagreb: Golden Marketing.

Livingstone Smith, David, ed. 2017. *How Biology Shapes Philosophy: New Foundations for Naturalism*. Cambridge: Cambridge University Press.

McGinn, Colin. 1976. "On the Necessity of Origin". *Journal of Philosophy*, 73: 127–35.

Machery, E., R. Mallon, S. Nichols, and S. P. Stich. 2004. "Semantics, Cross-Cultural Style". *Cognition*, 92(3): 1–12. https://doi.org/10.1016/j.cognition.2003.10.003.

Machery, E., C. Y. Olivola, and M. de Blanc. 2009. "Linguistic and Metalinguistic Intuitions in the Philosophy of Language". *Analysis*, 69(4): 689–94. https://doi.org/10.1093/analys/anp095.

Mackie, Penelope. 2006. *How Things Might Have Been: Individuals, Kinds, and Essential Properties*. Oxford: Clarendon Press.

Maglo, K. 2010. "Genomics and the Conundrum of Race: Some Epistemic and Ethical Considerations". *Perspectives in Biology and Medicine*, 53(3): 357–72.

Maglo, K. 2011. "The Case against Biological Realism about Race: From Darwin to the Post-Genomic Era". *Perspectives on Science*, 19(4): 361–90.

Mallet, James. 1995. "A Species Definition for the Modern Synthesis". *Trends in Ecology and Evolution*, 10: 294–9.

Mallon, Ron. 2018. "Constructing Race: Racialization, Causal Effects, or Both?" *Philosophical Studies*, 175: 1039–56. https://doi.org/10.1007/s11098-018-1069-8.

Mallon, Ron. Forthcoming. "Racial Essentialism". In Kathrin Koslicki and Michael Raven (eds.), *Routledge Handbook of Essence*. New York: Routledge.

Martí, Genoveva. 2009. "Against Semantic Multi-Culturalism". *Analysis*, 69(1): 42–8. https://doi.org/10.1093/analys/ann007.

Martí, Genoveva. 2012. "Empirical Data and the Theory of Reference". In W. P. Kabasenche, M. O'Rourke, and M. H. Slater (eds.), *Reference and Referring: Topics in Contemporary Philosophy*. Cambridge, MA: MIT Press: 62–76.

Martí, Genoveva. 2014. "Reference and Experimental Semantics". In E. Machery and E. O'Neill (eds.), *Current Controversies in Experimental Philosophy*. New York: Routledge: 17–26.

Matthen, Mohan. 1998. "Biological Universals and the Nature of Fear". *Journal of Philosophy*, 95: 105–32.

Mayr, Ernst. 1961. "Cause and Effect in Biology". *Science*, 134: 1501–6.

Mayr, Ernst. 1963. "Species Concepts and Their Application". From Chapter 2 of *Populations, Species, and Evolution*. Cambridge, MA: Harvard University Press. Reprinted in Ereshefsky (1992a), *The Units of Evolution: Essays on the Nature of Species*. Cambridge, MA: MIT Press: 15–25. [Citations are to Ereshefsky.]

Mayr, Ernst. 1969. *Principles of Systematic Zoology*. New York: McGraw-Hill.

Mayr, Ernst. 1982. *The Growth of Biological Thought*. Cambridge, MA: Harvard University Press.

Mayr, Ernst. 2002. "The Biology of Race and the Concept of Equality". *Daedalus*, 131(1): 89–94.

Millikan, Ruth Garrett. 1999. "Historical Kinds and the Special Sciences". *Philosophical Studies*, 95: 45–65.

Millikan, Ruth Garrett. 2000. *On Clear and Confused Ideas*. Cambridge: Cambridge University Press.

Mishler, Brent D. 1999. "Getting Rid of Species?" In Robert A. Wilson (ed.), *Species: New Interdisciplinary Essays*. Cambridge, MA: MIT Press: 307–15.

Mishler, Brent D., and Michael J. Donoghue. 1982. "Species Concepts: A Case for Pluralism". *Systematic Zoology*, 31: 491–503. Reprinted in Ereshefsky (1992a), *The Units of Evolution: Essays on the Nature of Species*. Cambridge, MA: MIT Press: 121–37.

Neander, Karen. 1996. "Swampman Meets Swampcow". *Mind & Language*, 11(1): 118–29.

Needham, Paul. 2011. "Microessentialism: What is the Argument?" *Noûs*, 45: 1–21.

Newman, Stuart A. 2020. "The Origins and Evolution of Animal Identity". In A. S. Meincke and J. Dupré (eds.), *Biological Identity: Perspectives from Metaphysics and the Philosophy of Biology*. New York. Routledge: 128–48.

Nichols, S., N. A. Pinillos, and R. Mallon. 2016. "Ambiguous Reference". *Mind*, 125(497): 145–75. https://doi.org/10.1093/mind/fzv196.

Nilsson, D. E., and S. Pelger. 1994. "A Pessimistic Estimate of the Time Required for an Eye to Evolve". *Proceedings of the Royal Society of London Series B: Biological Sciences*, 256: 53–8.

Novembre, John, Toby Johnson, Katarzyna Bryc, Zoltán Kutalik, Adam R. Boyko, Adam Auton, Amit Indap, Karen S. King, Sven Bergmann, Matthew R. Nelson, Matthew Stephens, and Carlos D. Bustamante. 2008. "Genes Mirror Geography within Europe". *Nature*, 456: 98–101.

Oderberg, David S. 2007. *Real Essentialism*. New York: Routledge.

O'Hara, Robert J. 1993. "Systematic Generalization, Historical Fate, and the Species Problem". *Systematic Biology*, 42: 231–46.

Okasha, Samir. 2002. "Darwinian Metaphysics: Species and the Question of Essentialism". *Synthese*, 131: 191–213.

Pedroso, M. 2012. "Essentialism, History, and Biological Taxa". *Studies in History and Philosophy of Biological and Biomedical Sciences*, 43: 182–90.

Pfenninger, Markus, and Klaus Schwenk. 2007. "Cryptic Animal Species are Homogeneously Distributed among Taxa and Biogeographical Regions". *BMC Evolutionary Biology*, 7: 121.

Pigliucci, Massimo. 2013. "What Are We to Make of the Concept of Race? Thoughts of a Philosopher–Scientist". *Studies in History and Philosophy of Biological and Biomedical Sciences*, 44: 272–7.

Pigliucci, Massimo, and Jonathan Kaplan. 2003. "On the Concept of Biological Race and Its Applicability to Humans". *Philosophy of Science*, 70(5): 1161–72.

Popper, Karl R. 1950. *The Open Society and its Enemies*. Princeton: Princeton University Press.

Preston, B. 2009. "Philosophical Theories of Artifact Function". In A. Meijers (ed.), *Philosophy of Technology and Engineering Sciences*. Amsterdam: Elsevier: 213–33.

Pušić, Bruno, Pavel Gregorić, and Damjan Franjević. 2017. "What do Biologists Make of the Species Problem?" *Acta Biotheoretica*, 65(3): 179–209. https://doi.org/10.1007/s10441-017-9311-x.

Putnam, Hilary. 1975. *Mind, Language and Reality*. Cambridge: Cambridge University Press.

Putnam, Hilary. 2001. "Reply to Michael Devitt". *Revue Internationale de Philosophie*, 208: 495–502.

Quine, Willard V. 1960. *Word and Object*. Cambridge, MA: MIT Press.

Quine, Willard V. 1961. *From a Logical Point of View* 2nd edn. Cambridge, MA: Harvard University Press.

Reichenbach, H. 1947. *Elements of Logic*. London: Macmillan.

Richards, Richard A. 2010. *The Species Problem: A Philosophical Analysis*. Cambridge: Cambridge University Press.

Robertson, T. 1998. "Possibilities and the Arguments for Origin Essentialism". *Mind*, 107: 729–49.

Robertson, T., and P. Atkins. 2016. "Essential vs Accidental Properties". In E. Zalta (ed.), *The Stanford Encyclopedia of Philosophy* (Summer 2016 Edition). https://platostanfordedu/archives/sum2016/entries/essential-accidental/.

Roca, A. L., N. Georgiadis, J. Pecon-Slattery, and S. J. O'Brien. 2001. "Genetic Evidence for Two Species of Elephant in Africa". *Science*, 293(5534): 1473–7. https://doi.org/10.1126/science.1059936.PMID 11520983.

Root, Michael. 2003. "The Use of Race in Medicine as a Proxy for Genetic Differences". *Philosophy of Science*, 70(5): 1173–83.

Rosenberg, Alex. 1985. *The Structure of Biological Science*. New York: Cambridge University Press.

Rosenberg, Noah A. 2011. "A Population-Genetic Perspective on the Similarities and Differences among Worldwide Human Populations". *Human Biology*, 83(6): 659–84.

Rosenberg, Noah A., Jonathan K. Pritchard, James L. Weber, Howard M. Cann, Kenneth K. Kidd, Lev A. Zhivotovsky, and Marcus W. Feldman. 2002. "Genetic Structure of Human Populations". *Science*, 298(5602): 2381–5.

Ruhl, Charles. 1989. *On Monosemy: A Study in Linguistics Semantics*. Albany: SUNY Press.

Ruse, Michael. 1987. "Biological Species: Natural Kinds, Individuals, or What?" *British Journal for the Philosophy of Science*, 38: 225–42. Reprinted in Ereshefsky (1992a), *The*

Units of Evolution: Essays on the Nature of Species. Cambridge, MA: MIT Press: 343–61. [Citations are to Ereshefsky.]

Salazar-Ciudad, Isaac, Richard Sole, and Stuart A. Newman. 2001. "Phenotypic and Dynamical Transitions in Model Genetic Networks I: Emergence of Patterns and Genotype-Phenotype Relations". *Evolution and Development*, 3: 84–94.

Salmon, Nathan. 1979. "How Not to Derive Essentialism". *Journal of Philosophy*, 76: 703–25.

Salmon, Nathan. 2020. "Naming and Non-Necessity". In Andrea Bianchi (ed.), *Language and Reality from a Naturalistic Perspective: Themes from Michael Devitt.* Cham: Springer: 237–48.

Sartorelli, J. 2016. "Biological Process, Essential Origin, and Identity". *Philosophical Studies*, 173: 1603–19.

Schiffer, Stephen. 1978. "The Basis of Reference". *Erkenntnis*, 13: 171–206.

Schwartz, S. P. 1978. "Putnam on Artifacts". *Philosophical Review*, 87: 566–74.

Schwartz, S. P. 2002. "Kinds, General Terms, and Rigidity". *Philosophical Studies*, 109: 265–77.

Searle, John R. 1983. *Intentionality: An Essay in the Philosophy of Mind.* Cambridge: Cambridge University Press.

Sesardic, Neven. 2010. "Race: A Social Destruction of a Biological Concept". *Biology and Philosophy*, 25(2): 143–62.

Simpson, George G. 1945. "The Principles of Classification and a Classification of Mammals". *Bulletin of the American Museum of Natural History*, 85: 1–350.

Simpson, George G. 1961. *Principles of Animal Taxonomy.* New York: Columbia University Press.

Simpson, Michael G. 2010. *Plant Systematics.* 2nd edn. Burlington, MA: Academic Press.

Slater, Matthew H. 2013. *Are Species Real? An Essay on the Metaphysics of Species.* New York: Palgrave Macmillan.

Smith, H. M. 1999. "Comment on the Proposed Conservation of *Coluber infernalis* Blainville, 1835 and *Eutaenia sirtalis tetrataenia* Cope in Yarrow, 1875 (Currently *Thamnophis sirtalis infernalis* and *T. s. tetrataenia*; Reptilia, Squamata): Proposed Conservation of the Subspecific Names by the Designation of a Neotype for *T. s. infernalis*". *Bulletin of Zoological Nomenclature*, 56(1): 71–2.

Smith, H. M., D. Chiszar, and R. R. Montanucci. 1997. "Subspecies and Classification". *Herpetological Review*, 28: 13–16.

Sober, Elliott. 1980. "Evolution, Population Thinking and Essentialism". *Philosophy of Science*, 47: 350–83. Reprinted in Ereshefsky (1992a), *The Units of Evolution: Essays on the Nature of Species.* Cambridge, MA: MIT Press: 247–78. [Citations are to Ereshefsky.]

Sober, Elliott. 1984. "Sets, Species and Evolution: Comments on Philip Kitcher's 'Species' ". *Philosophy of Science*, 51(2): 334–41.

Sober, Elliott. 1993. *Philosophy of Biology.* Boulder: Westview Press.

Sokal, Robert, and Theodore J. Crovello. 1970. "The Biological Species Concept: A Critical Evaluation". *American Naturalist*, 104: 127–53. Reprinted in Ereshefsky (1992a), *The Units of Evolution: Essays on the Nature of Species.* Cambridge, MA: MIT Press: 27–55. [Citations are to Ereshefsky.]

Spencer, Quayshawn. 2012. "What 'Biological Racial Realism' Should Mean". *Philosophical Studies*, 159: 181–204. https://doi.org/10.1007/s11098-011-9697-2.

Spencer, Quayshawn. 2014. "The Unnatural Racial Naturalism". *Studies in History and Philosophy of Biological and Biomedical Sciences*, 46: 38–43.

Spencer, Quayshawn. 2018a. "Racial Realism I: Are Biological Races Real?" *Philosophical Compass*, 13(1): e12467.
Spencer, Quayshawn. 2018b. "Racial Realism II: Are Folk Races Real?" *Philosophical Compass*, 13(1): e12467.
Spencer, Quayshawn. 2019. "How to Be a Biological Racial Realist". In Joshua Glasgow, Sally Haslanger, Chike Jeffers, and Quayshawn Spencer (eds.), *What Is Race? Four Philosophical Views*. Oxford: Oxford University Press: 73–110.
Stanford, P. Kyle. 1995. "For Pluralism and Against Realism about Species". *Philosophy of Science*, 62: 70–91.
Sterelny, Kim. 1999. "Species as Ecological Mosaics". In Robert A. Wilson (ed.), *Species: New Interdisciplinary Essays*. Cambridge, MA: MIT Press: 119–38.
Sterelny, Kim, and Paul Griffiths. 1999. *Sex and Death*. Chicago: University of Chicago Press.
Strotz, L. C., and A. P. Allen. 2013. "Assessing the Role of Cladogenesis in Macroevolution by Integrating Fossil and Molecular Evidence". *Proceedings of the National Academy of Sciences*, 110(8): 2904–9.
Sullivan, Arthur. 2010. "Millian Externalism". In Robin Jeshion (ed.), *New Essays on Singular Thought*. Oxford: Oxford University Press: 246–69.
Sundstrom, Ronald R. 2002. "Race as a Human Kind". *Philosophy and Social Criticism*, 28 (1): 91–115.
Szathmary, Eors. 2002. "Evolution: Developmental Circuits Rewired". *Nature*, 411: 143–5.
Tahko, Tuomas E. 2015. "Natural Kind Essentialism Revisited". *Mind*, 124: 795–822.
Taylor, P., M. Lopez, J. Martínez, and G. Velasco. 2012. "When Labels Don't Fit: Hispanics and their Views of Identity". Washington, DC: The Pew Hispanic Center.
Templeton, Alan R. 1999. "Human Races: A Genetic and Evolutionary Perspective". *American Anthropologist*, 100(3): 632–50.
Templeton, Alan R. 2013. "Biological Races in Humans". *Studies in History and Philosophy of Biological and Biomedical Sciences*, 44(3), 262–71.
Thompson, Michael. 2008. *Life and Action: Elementary Structures of Practice and Practical Thought*. Cambridge, MA: Harvard University Press.
Turland, N. J., J. H. Wiersema, F. R. Barrie, W. Greuter, D. L. Hawksworth, P. S. Herendeen, S. Knapp, W.-H. Kusber, D.-Z. Li, K. Marhold, T. W. May, J. McNeill, A. M. Monro, J. Prado, M. J. Price, and G. F. Smith, eds. 2018. *International Code of Nomenclature for Algae, Fungi, and Plants (Shenzhen Code) adopted by the Nineteenth International Botanical Congress Shenzhen, China, July 2017*. Regnum Vegetabile 159. Glashütten: Koeltz Botanical Books. https://doi.org/10.12705/Code.2018.
Valentine, J. W. 1995. "Why No New Phyla after the Cambrian? Genome and Ecospace Hypotheses Revisited". *Palaios*, 5(10): 190–4.
Van Valen, Leigh. 1976. "Ecological Species, Multi-Species and Oaks". *Taxon*, 25: 233–9. Reprinted in Ereshefsky (1992a), *The Units of Evolution: Essays on the Nature of Species*. Cambridge, MA: MIT Press: 69–77. [Citations are to Ereshefsky.]
Wade, Nicholas. 2006. "Scientists Hope to Unravel Neanderthal DNA and Human Mysteries". *New York Times*, July 21: A14.
Walsh, Denis. 2006. "Evolutionary Essentialism". *British Journal for the Philosophy of Science*, 57: 425–48.
Webster, Gerry, and Brian C. Goodwin. 1996. *Form and Transformation: Generative and Relational Principles in Biology*. Cambridge: Cambridge University Press.
Wettstein, Howard. 2012. "On Referents and Reference Fixing". In R. Schantz (ed.), *Prospects for Meaning*. Berlin: De Gruyter: 107–18.

Wiggins, David. 1980. *Sameness and Substance*. Oxford: Blackwell.
Wiley, E. O. 1981. *Phylogenetics: The Theory and Practice of Phylogenetic Systematics*. New York: John Wiley.
Wilkins, John S. 2010. "What is a Species? Essences and Generation". *Theory in Biosciences*, 129: 141–8.
Wilson, Robert A., ed. 1999a. *Species: New Interdisciplinary Essays*. Cambridge, MA: MIT Press.
Wilson, Robert A. 1999b. "Realism, Essence, and Kind: Resuscitating Species Essentialism?" In Wilson (1999a), *Species: New Interdisciplinary Essays*. Cambridge, MA: MIT Press: 187–207.
Wilson, Robert A., Matthew J. Barker, and Ingo Brigandt. 2007. "When Traditional Essentialism Fails: Biological Natural Kinds". *Philosophical Topics*, 35: 189–215.
Winsor, Mary P. 2006. "The Creation of the Essentialism Story: An Exercise in Metahistory". *History and Philosophy of the Life Sciences*, 28: 149–84.
Witteveen, J. 2015. "Naming and Contingency: The Type Method of Biological Taxonomy". *Biology and Philosophy*, 30: 569–86. https://doi.org/101007/s10539-014-9459-6.
Wyatt, Sharon B., David R. Williams, Rose Calvin, Frances C. Henderson, Evelyn R. Walker, and Karen Winters. 2003 "Racism and Cardiovascular Disease in African Americans". *The American Journal of the Medical Sciences*, 325(6): 315–31.

Index

For the benefit of digital users, indexed terms that span two pages (e.g., 52–53) may, on occasion, appear on only one of those pages.